Automating Your Office

G L Simons

PUBLISHED BY NCC PUBLICATIONS

British Library Cataloguing in Publication Data

Simons, G. L.
Automating your office.
1. Office practice – automation
2. Office management
I. Title II. National Computing Centre
651 HF5548.2

ISBN 0-85012-402-6

First published in 1984 by:

NCC Publications, The National Computing Centre Limited, Oxford Road, Manchester M1 7ED, England.

Typeset in 11pt Times Roman by UPS Blackburn Limited, 76-80 Northgate, Blackburn, Lancashire, and printed by Hobbs the Printers of Southampton.

ISBN 0-85012-402-6

Acknowledgements

The information in this publication is largely derived from articles, books and discussions with colleagues. I am indebted to various authors and co-authors of NCC books (see Appendix 1):

I R Beaman
G B Bleazard
I Cole
M A Condon
L Damodaran
A Elbra
R J Firth
K C E Gee
M Peltu
S G Price
J A T Pritchard
A Simpson
P A Wilson
M B Wood

I am also grateful to Linda Barrett for making editorial suggestions and for checking the material at various stages.

Geoff Simons

Contents

	Page
Acknowledgements	
PART 1 BACKGROUND	11
1 The Electronic Office	13
Introduction	13
Information as a Resource	14
The Automation Approach	15
Benefits of Office Automation	17
Functions and Equipment	18
Summary	21
2 Office Automation Facilities	23
Introduction	23
Word Processing	24
Electronic Mail	25
Conferencing	27
Information Access	28
Facsimile	29
Networks	31
Summary	33
3 A Strategy for Change	35
Introduction	35

The Strategic Framework 37
Diagnosis: Objectives and Targets 38
Design: Staffing and Costs 40
Development: Systems and Suppliers 42
Transition: Managing Change 43
Implementation: the Working System 44
Adaptation: Learning from Experience 45
Summary 46

PART 2 TOWARDS OFFICE SYSTEMS 49

4 The Office in Transition 51

Introduction 51
Serving the Customer 52
Managing the Organisation 53
The Management Structure 54
Control within the Organisation 55
The Role of Middle Managers 56
Informal Factors 57
Organising Office Work 58
Office Information Services 59
Information Management 59
The Office Technology Project 61
Summary 62

5 Approaching Systems Design 63

Introduction 63
The Feasibility Study 64
Managing the Feasibility Study 64
Terms of Reference 65
Topics to be Studied 65
Presenting the Study Conclusions 66
Elements in Systems Design 67
Design for the User 69
Workplace Design 71
Summary 72

6 Choosing Office Systems 73

Introduction 73
Cost Factors 75
Vendors 77
Going to Tender 79
Evaluating the Proposals 81
Agreeing a Contract 83
Summary 85

PART 3 THE HUMAN FACTOR 87

7 Managing People in the Office 89

Introduction 89
Personnel Management 90
The Employment Question 90
Technology Agreements 92
Motivation and Morale 96
The Need for Training 97
Secretarial, Typing and Clerical Skills 98
Management Skills 99
Summary 100

8 People and Health 103

Introduction 103
Health and Safety 104
Visual Factors 105
Postural Factors 108
Ergonomics 108
Summary 110

PART 4 THE FUTURE 113

9 Tomorrow's Office 115

Introduction 115
Influential Factors 116
The Integrated System 117
Automation Milestones 118
Expert (Office) Systems 119
Summary 121

Appendix

1 Bibliography 123

2 Glossary 127

3 Sample Tender Document 143

4 Sample Letter to Suppliers 153

5 Cost Elements 157

6 Checklists 161

Index 199

Part 1
Background

1 The Electronic Office

INTRODUCTION

The impact of computer-based systems is being felt in all areas of society. An expanding range of new equipment is being made available for factories and commerce, education and medicine, defence and communications, entertainment and the domestic environment. Computer-based facilities are increasingly evident in offices, and this trend is likely to develop rapidly in the future.

Computers were first used in offices in the 1950s as efforts were made to automate various accounting and record-keeping functions. In fact one of the early commercial computers, developed by the British catering company Lyons, was called the Lyons Electronic Office (LEO), so anticipating by two decades the later interest in the *electronic office*. But we should remember that the early commercial computers, physically massive by today's standards, were invariably sited in special computer centres remote from actual office environments. It was not until the 1970s that microelectronics was able to bring computer power into the office.

Towards the end of the 1970s there was increasing talk of the *electronic office* – or the *office of the future,* the *'paperless' office,* the *automated office*, etc. The various phrases were intended to denote the same broad concept, though technology was moving so fast that it was difficult to describe the emerging possibilities fully and without ambiguity. What *was* clear was that offices were being influenced – in terms of their equipment, staffing, operational structures, etc – by new developments in microelectronics, just as the emerging technology was influencing other social sectors.

The electronic office, often depicted as the future realisation of various computer-based possibilities, can be conceived in radical terms. It is often portrayed as being completely automated, though this suggestion conjures up a misleading analogy with factory automation (see The Automation Approach, below). The electronic office is also seen as 'paperless' (there is much talk of 'paper pollution' in the conventional office), and as representing a 'function' or a 'concept' rather than a fixed geographical entity. The office is traditionally a place in a building, to which people travel and in which they are expected to remain for well-defined periods. But computer-based facilities can now enable geographically-dispersed individuals to perform various office tasks on a co-operative basis. The development of microelectronics is stimulating a radical reappraisal of the concept of the office.

We can see that the electronic office will manifest itself in various ways, according to the requirements of the organisation and the needs of the individual users. It is likely that linked workstations will embody many office facilities (see Chapter 2), each workstation resembling a screen-based stand-alone terminal. There will be flexibility in the performance scope of individual terminals, in the provision of back-up storage, in the siting of particular units. The overall objective is to provide the necessary office services with maximum efficiency at minimum cost.

INFORMATION AS A RESOURCE

There is growing recognition of the importance of information. The individual needs information to perform a task; the company cannot survive without it; the country needs information for countless purposes. It is easy to see how information can be depicted as a *resource,* as essential to productive success as are raw materials and skilled staff. But information has to be collected, manipulated, *managed.* The significant contribution of electronic office technology will be to enhance the effective management of information in a wide range of office environments.

It is also important to appreciate the *economic* significance of information. Where company information is used effectively, the levels of trade and revenue are maximised. Moreover it costs money to acquire and generate information. Some information is so important that individual companies may be unwilling to divulge it at any price. And the problem of managing information for commercial and other purposes has been compounded by the sheer volume of information being made available.

As far back as the mid-1950s, according to a US report (*Records and Paperwork Management in the Federal Government: Two Decades of Recommendations,* cited by Pritchard and Cole, 1983), there were in excess of 25 billion cubic feet of government records in existence, with annual records management costing around $4 billion. In the subsequent decades this amount of documentation increased significantly, a development that was paralleled – though at lower volume levels – in thousands of companies and other organisations.

It became increasingly clear by the 1980s that traditional labour-intensive, paper-based, information-handling methods would – with the rapid growth in the number of documents – become progressively less efficient. The electronic office offered a solution, possibly the only realistic solution, to a worsening office situation that threatened the very survival of many companies in the 1980s. The slogan *'automate or die'* seemed increasingly relevant to offices faced with the pressing need to handle ever-larger volumes of information with ever-greater efficiency.

The recognition that information is a vital economic resource underlines the need for effective office systems. Office services invariably focus on the collection, processing or distribution of information. It is increasingly acknowledged that in all these tasks, computer-based facilities have a vital part to play. Office automation is essential to the commercial viability of companies in the 1980s and beyond.

THE AUTOMATION APPROACH

Automation is a generalised concept. It is concerned with the development and operation of systems that rely on little or no human intervention. In principle, any type of labour-intensive process can be automated, and this development has been particularly evident in manufacturing industry.

Efforts were made, long before the emergence of electronic

digital computers, to automate manufacturing processes. Industrialists such as Henry Ford realised that automation could dramatically boost the efficiency of production in the factory environment. The early manufacturing innovators were influenced by the work design techniques advocated by an American engineer, Frederick Winslow Taylor. His approach – known as *Taylorism* or *Scientific Management* – relied upon breaking work routines into small steps which could be repeatedly performed using a few simple skills. This allowed tasks performed by workers to be precisely defined, with rewards to workers based on achieving and exceeding defined targets.

The Taylor approach required that production workers perform the same simple tasks, time after time, with no scope for initiative or creative activity of any sort. Workers were required to behave like machines, a dehumanising circumstance vividly portrayed in the 1926 film *Metropolis*. This depiction also uses the familiar idea that an artificial creation may be superior to a human being. Hence the scientist/magician Rotwang declares: 'I have created a machine in the image of man, that never tires or makes a mistake . . . Now we have no further use for living workers'.

The application of Taylor methods to production prepared the way for automation: it became relatively easy – in such areas as spray-painting, welding and mechanical assembly – to replace workers by robots and other automated systems capable of simple, repetitive activity. Today there is much talk of the 'unmanned factory', and there are already production units in the US, Europe and Japan that can function without human intervention.

It is obvious that a similar automation approach to the office environment would have profound consequences for office staff (see Part 3, The Human Factor). But it should be emphasised that the current approach to the electronic office does not envisage an absence of human intervention. Computer-based office systems are seen as aids to human activity. They are relevant to questions of employment, health, motivation and de-skilling, but managers should remember that people are essential elements in office systems. If office automation is implemented in disregard of the people concerned, it will inevitably be counter-productive.

Office work, in contrast to industrial activity, has traditionally been relatively free and unstructured. Staff have, to some extent, been able to develop their own patterns of work, an evident flexibility that has helped organisations to remain responsive to change in the business environment. It has been emphasised (eg by Peltu, 1984) that an excessively Tayloristic bias in the implementation of office automation would have many adverse consequences. Office managers, in developing computer-based systems, should aim to encourage the human commitment without which the systems cannot operate efficiently.

BENEFITS OF OFFICE AUTOMATION

In recent years a number of factors have combined to accelerate the trend towards the electronic office: most of these factors relate to costs. The cost of much computer-based equipment has fallen, in relation to the facilities offered, whereas over a similar period wages, salaries, and the cost of office supplies (such as paper and furniture) have increased significantly. Office rental charges have also increased, as have the costs of transporting people to conferences and other meetings. There are obvious pressures on managements, in these circumstances, to reduce such costs without impairing business efficiency. Computer-based equipment can help managers in this task, though it seems clear that there will be inevitable consequences for staffing levels.

The main benefit that organisations are likely to derive from office automation will be the increased productivity of staff at all levels. In particular, it is hoped, higher levels of productivity will be found in managers, decision-makers, creative thinkers and in the sales force – all these are staff responsible directly for the commercial viability (survival, expansion, etc) of the company. This serves to emphasise that managers should not move towards the electronic office merely to improve the productivity of clerks, secretaries and typists. There is immense potential for productivity in computer-based office systems, and this relates to the work of the highest-paid, as well as to that of the lowest-paid, individuals in the organisation.

It should also be emphasised that the full benefits and savings of office automation systems will not be realised unless *fully integrated* systems are implemented. The greater the number of office workers who have access to a workstation linked to an office automation network, the more valuable the network to the organisation. It is important to achieve what has been called a 'critical mass' of users of the automated office system.

To summarise, the benefits of office automation are likely to derive from cutting costs, on the one hand, and increasing revenue, on the other. *Added-value applications* are seen as realising the full potential of office automation. Such applications bring four main types of benefits:

— *increased information accessibility,* to give information that is more current and more accurate. A wide range of people can be given access to the latest version;

— *increased people accessibility,* to permit people to communicate efficiently, even though they may be geographically separated or not simultaneously available. Time zones and conflicting activities can cease to be barriers to effective communication. Conferencing systems and file-sharing systems allow geographically-separate workers to work together, to enter discussions, and to take joint decisions;

— *increased control over personal activities,* gained through enabling the individual to work effectively from a variety of locations, including home and hotel rooms;

— *increased individual contributions,* because now the individual can spend more time on the primary task rather than on the related support activities. People have better access to up-to-date information and more time for crucial work.

The electronic office brings a range of productivity gains to all levels of staff, provided that the systems are implemented with intelligent regard to all the relevant factors. The productivity gains enhance the commercial viability of the company, allowing, for example, for diversification of activity and for expansion into new markets.

FUNCTIONS AND EQUIPMENT

In one typical breakdown (Price, 1979), electronic office technology is seen as relevant to four areas of activity:

— *document preparation,* including such activities as dictation, typing, updating of documents, duplicating, copying, and the use of typesetting services;

— *message distribution,* including telephoning, mail services, and travel for meetings;

— *personal information management,* including the use of local filing cabinets, in/out trays, diaries, planning charts; the filing of information that adds to a corporate data base;

— *information access,* including making references to dictionaries, timetables and catalogues, the use of library and bibliographic services, and access to corporate data bases.

The traditional office contains various familiar items, such as the desk, the typewriter, the telephone and the filing cabinet. These items facilitate such tasks as document generation, communica-

Information Activity	Traditional Methods	New Office Technologies
Document generation –letters, memos, reports, etc	Typewriter Pen/paper Printer	Word processor Electronic typewriter Computer printer
Communication –exchanging information –gathering information –disseminating information	Telephone Meetings Copier/facsimile Telex Mail Publishing	Enhanced telephony Teleconferencing Digital facsimile Teletex Electronic mail Videotex/viewdata Computer networks
Filing –storing information in a way that assists retrieval of items	Filing cabinet Library	Electronic filing Databases Information retrieval Videotex/viewdata Computerised storage

Figure 1.1 Key Office Activities and Technologies

tions and filing – to which office automation can be successfully applied (see Figure 1.1). Figure 1.2 shows how electronic systems can be applied to information processing, decision making and transaction processing – all typical office functions (eg involved in helping a client, deciding upon a course of action, selecting an option, preparing the accounts). Electronic systems can both duplicate traditional methods and extend their scope.

Office automation is relying upon the convergence of the technologies of computing, telecommunications and office systems. A wide range of computer-based equipment is now affecting methods of working in the office environment. The following types of equipment are being installed in offices alongside telephones, typewriters, photocopiers and telex machines:

— *word processors,* both as stand-alone devices or as units in larger integrated systems (involving, for example, other word processors, computers or phototypesetting equipment);

Functioned Performed	Some Tasks Involved	Electronic Systems Used
Information processing	Accounts processing Payroll processing Stock control Calculations	Data Processing (DP) systems Personal (micro) computers Software packages
Decision making	Gathering information Analysing information Predicting result of possible decisions	Decision Support Systems (DSS) Modelling software Electronic spreadsheets Databases Information retrieval Electronic mail
Transaction processing	Dealing with an order Answering a query Making a seat booking	Combination of new office technologies often including direct user links to computer system via a telecommunications link

Figure 1.2 Office Functions Using Electronic Systems

— *small business computers,* involved in accounting, word processing and other functions;

— *facsimile equipment* for document transmission between, say, the various branches of a company;

— *viewdata* sets for interacting with the British Telecom Prestel system or with private viewdata systems;

— *terminals* linked to computers for various purposes, such as providing immediate access to up-to-date information. Intelligent terminals, incorporating their own computer power, can perform many business functions (eg financial modelling, market analyses, economic projections).

Some of these items of equipment may be installed in homes, linked to company systems via the telephone network. Another option is for portable workstations to enable people to work while travelling. Such provisions show that the electronic office need not be physically restricted to a particular geographical site. The flexibility of office automation is well illustrated by the fact that portable workstations can aid travel where appropriate and teleconferencing systems can make travel unnecessary in some circumstances.

SUMMARY

This chapter has outlined features of the electronic office. Some background details have been included, and attention has been given to the concept of automation and to the idea that information is an economic resource. Benefits of office automation are indicated, with details of typical office activities and electronic systems.

Already some guidelines for implementing electronic office systems have emerged. For example, it is important to note that:

— attention to people factors is important to the successful implementation of electronic office systems;

— the benefits of office automation are relevant to the work of staff at all levels, not just to the activities of clerks and typists;

— it is important to aim for fully integrated office systems, instead of allowing separate computer-based facilities to develop in an unco-ordinated piecemeal fashion;

— office automation should be adapted to a company's needs in a flexible way. A thoughtful management strategy is essential for this purpose.

These topics, and others, will be explored in the chapters that follow.

2 Office Automation Facilities

INTRODUCTION

Today many available commercial products belong in the electronic office. Often, however, it is difficult to link these in effectively integrated systems. Some of the products and facilities, such as Telex and word processors, have existed for many years whereas others, such as electronic mail and Prestel, are more recent and generate much speculation about their potential for the future. Yet other facilities, such as the private branch exchanges, have evolved over more than a decade, yet continue to show possibilities for future development. The stored-program controlled telephone exchange can provide a wide range of sophisticated voice handling, data switching and environmental control facilities (handsets connected to a PBX can now be viewed as useful data terminals facilitating the collection of data and the making of enquiries).

This chapter introduces the reader to the more common office-automation facilities. Not all the available facilities are mentioned but, where possible, reference is made to publications that give more information. A main purpose of the chapter is to show that the electronic office can contain many disparate elements, variously concerned with the collection, storing, handling and dissemination of information. For the various facilities to operate in a co-operative, mutually supportive way, it is necessary that they be linked in a networking system: mention is made of local area networks as a means to this end.

Again the manager contemplating the implementation of electronic office systems is advised to adapt the available products to the needs of the particular company or organisation. The manager should first become acquainted with the broad spectrum of office-automation facilities; then a strategy for change (see Chapter 3) should be developed.

WORD PROCESSING

Apart from such traditional facilities as Telex and the telephone, the word processor is the most familiar element in the electronic office. In recent years it has received more attention than most other facilities associated with office automation. It is easy to see why this should be so. The word processor can easily be introduced as a straightforward enhancement to the typing function, and benefits can be quickly quantified. For example, NCC's 1981 Multi Client Study (see Bleazard and Price) observes that automation of the typing function had produced significant improvements in productivity. (Wilson and Pritchard, 1983, report productivity changes, following the introduction of word processors, in various companies.)

Managers should not rush to acquire word processing systems without first assessing in detail the company's (or department's) current needs, likely future requirement, and the available electronic office systems on the market. It may be, for example, that a business system with a word processing capability may be a better option than a simple word processor with no capacity for any other type of operation.

Word processors offer a wide range of facilities (see, for example, Simons, 1981, Chapter 2). These fall into four broad categories:

- speeding up the initial text keying and reducing problems of text layout;
- modifying text recorded on magnetic storage;
- using stored text together with newly-keyed text to create a document;
- improving printed text quality and automating the various paper-handling functions.

In addition to these direct text-processing functions, many available word processors have evolved to embrace traditional data processing functions and to link other components in the electronic office. This should be remembered when managers are planning the acquisition of word-processing equipment. By the late-1970s, commercial systems were offering multiple facilities: for instance, the J100 Videocomputer from Jacquard was launched as a simultaneous data-processing, word-processing and communications system. In fact the screen-based word processing system and the small data processing system are likely to use very similar hardware, a circumstance that has made it easy to integrate WP and DP functions.

Many of the quickly perceived benefits of word processing are tangible and obvious, relating as they do to typist productivity, work throughput, and the quality of finished documents. Other benefits, less tangible, are equally important. These relate to the job satisfaction of WP operators and to improved customer/client goodwill due to the faster processing of correspondence. But any realised benefits will only be evident where a sound implementation strategy has been developed. A small company may only require a modest investment in word processing and other office-automation equipment whereas a large company may expect to spend hundreds of thousands of pounds developing all available electronic office systems in massive integrated systems. Whatever the case, a coherent management strategy, carefully adapted to the organisation's needs, is essential.

ELECTRONIC MAIL

Traditional mailboxes – whether found in the home, the office, etc – allow people to send and receive messages. A conventional mailbox, free-standing or set in a door or wall, provides a message delivery and access point which people use to aid the task of communication. Today we are witnessing the emergence of the *electronic* mailbox, where the traditional concept is being exploited as a key element in information technology. This development is of importance, not only to the electronic office, but also to factories, schools, hospitals, the domestic environment – wherever people need to send and receive communications.

Electronic mailboxes are run using computers and terminals.

Mailboxes are currently available on viewdata computers which can be accessed using television receivers as terminals. Software is used to organise the communication and information flows: for example, to switch the messages between the mailboxes. One possibility is that mailboxes will become associated with distributed networks of small computers, or even with individual microcomputers (see Wilson, 1984).

Various factors are combining to encourage the development of the electronic mailbox. The costs of paper-based mail have rapidly increased in recent years; electronic mail transmission costs have been found to compare favourably with charges for special rapid delivery services; many companies are becoming dissatisfied with the time taken to deliver paper-based mail; and there are advantages in being able to integrate electronic mail into systems with computer-based data processing, filing and retrieval systems.

In some cases, electronic mailbox provisions have been superimposed onto the general data communications network. For example, for several years Texas Instruments has operated a distributed minicomputer network spanning four continents and including 52 distributed computing nodes. In addition to the corporate data processing facility, there is an electronic mail system which can route messages to any terminal or printer on the network: by 1980, around 21,000 messages a day were being handled via the mail facility.

It is envisaged that electronic mail systems will be key elements in integrated electronic office systems: for example, communicating word processors could be linked to office terminals via electronic mail facilities. We may expect all office workstations to have access to a mail system which will allow communication with all other terminals inside the organisation via a site message switch machine. Arrangements should be made to protect the privacy of items being transmitted via the mailbox facility: any one person should only be able to access one mailbox.

As with word processors and other elements in the electronic office, the manager should view the mailbox as a likely component in an integrated system suited to the needs of the organisation. The experience of various companies with their electronic mailbox systems is well surveyed in Wilson and Pritchard (1983, pp 57-64).

CONFERENCING

All electronic conferencing systems aim to reduce the travel burden on individuals: conferencing may be seen as a substitution for travel. In some cases an individual may still need to travel to an important meeting but can use a conferencing system to communicate (be 'present') with colleagues back at the office. In this way an individual can, in effect, be in more than one place at the same time. Travel can be scheduled on a productive basis, and sometimes avoided altogether without detriment to communication efficiency. Some conferencing systems are linked to data processing and other facilities.

There are various types of (tele)conferencing systems:

— voice conferencing;

— video conferencing (continuous or snapshot);

— computer conferencing (screen sharing);

— computer conferencing (file sharing).

Many modern private branch exchanges offer voice conferencing facilities through existing telephone headsets. It is also possible to rent audio conference facilities and to establish good-quality sound transmission between groups of people. Voice conferencing offers some personal contact but, compared to a 'real' conference, is limited in scope.

By 1980 the British Post Office was offering the Confravision video conferencing system, allowing both sound and continuous television pictures. Mobile studios can be set up on company premises to facilitate regular use of the Confravision system. A simpler system is snapshot video, in which single-frame pictures are transmitted as required (this avoids the need for high-speed transmission lines). Documents and charts can be displayed but there is less personal contact than with continuous video systems.

Computer conferencing can utilise shared terminal screen images and be supplemented by voice conferencing. The conference members can each have a terminal screen of text/data which can be amended and discussed as required. Such a facility is available in the Tymshare AUGMENT system. Here users have access to the processing power of computers, copies of screens can be held in various states, information bases can be accessed, and contributions can be developed in private using text editing facilities.

Full video conferencing is likely to be regarded by many companies as too expensive. And the Confravision service, involving in some cases having to organise groups of people to attend city-centre studios, has been seen as inconvenient by some users. It has been suggested (eg by Pritchard and Wilson, 1982) that, because of the likely cost involved, only senior management workstations will have a video conferencing facility, with the workstations of other staff supporting only text, graphics, voice and data. One problem is that full video conferencing requires high-speed, high-capacity transmission lines, beyond the scope of many companies.

The various conferencing facilities bring a number of obvious benefits (as well as the loss of face-to-face contact) to companies, but the more expensive systems are unlikely to become very widespread. No system should be implemented as merely an attractive gimmick. The manager needs to be sure that any proposed conferencing system, as an element in his department's (or company's) integrated electronic office configuration, will make a demonstrable contribution to productivity, ie it will help the firm to become more commercially viable.

INFORMATION ACCESS

The integrated electronic office system has access to the various corporate data bases in the organisation. Staff can draw on up-to-date information to help them in their day-to-day tasks. Furthermore, workstations can be linked, via networking, to external information bases that cover mostly scientific, technical and bibliographic material.

In the UK there are a number of 'information supply' services; for example: the British Library Automated Information Service (BLAISE); the European Space Agency – Space Documentation Service, the General Electric Information System, and the Systems Development Corporation Search Service. Users can access well over a hundred on-line information bases covering a wide range of specialist topics (eg electrical components, chemical research, and developments in design).

Terminals on dedicated lines can be used to access the data bases. This arrangement is particularly useful where the volume of enquiries is high. It is also possible to use terminals connected by dialling over the PSTN.

There is already an evident growth in the use of on-line data bases, suggesting that the capacity to access such information will be provided as a typical facility on the electronic office workstations of technical and professional workers. Systems such as Prestel are acquainting users with the advantages of being able to draw on large volumes of relevant material in a short time. Information bases are growing rapidly, and also being made more easily accessible via networking.

It is obvious that managers should be aware of the types of specialist information that are required by their company. They should consider whether the workstations in their integrated office should have the facility to access external, as well as internal, collections of information. It may be, for example, that only one workstation should be linked to an external data base via a dedicated line, the remaining workstations being able to access the internal corporate data bases of the organisation.

FACSIMILE

Facsimile systems have been working successfully for many decades as the most effective way of transmitting non-codeable images. Obvious uses of facsimile are in transmitting weather maps or newspaper photographs from one town or country to another. Mobile facsimile systems are being used in police cars, fire engines, ships and offshore oil rigs. The vast majority of large companies, in the US and elsewhere, use facsimile systems in one form or another for business applications.

In facsimile systems, an original (chart, diagram, photograph, etc) is electronically scanned to produce electrical signals which can be transmitted to one or more locations where reproductions of the original can be generated. The facility provides fast delivery, highly authentic reproduction, and labour savings.

Japan has shown great initiative in developing facsimile systems, partly because of the complexities of the Japanese language (ie it is not easy to code Japanese characters). Facsimile units are much more common in Japan – when, for example, related to the number of telephones – than in Europe or the US. Pritchard and Cole (1983) cite Japanese equipment combining facsimile and other provisions. For example, the On-line Data Entry Facsimile System (ODEFAS) includes the capacity for optical character recognition (OCR) and information processing; and Toshiba is developing an integrated facsimile/OCR system.

Facsimile can be used for the inter-site intra-organisation distribution of documents (such as brochures received from other organisations, plans, drawings and sketches). And it can also be useful to transmit documents of various types from one organisation to another (eg between a company and its advertising agency, and vice versa). One obvious advantage of facsimile is the speed with which a document can be made available to geographically-distant people. This can allow, for instance, a number of interested parties to agree on a complex document before an imminent deadline.

It has been suggested (Pritchard and Wilson, 1982) that the technologies of facsimile, OCR and photocopying will lead to the 'image scanner'; and that facsimile, graph plotting and photcopying will lead to the 'image printer'. Such units could be installed as central facilities, as are today's photocopiers. Most of the information needed by workers will be available via their workstations, but the centralised image printer/scanner facility will be required to feed hard-copy information into the electronic system.

As with other elements of the electronic office, facsimile needs to be evaluated against the specific needs of the organisation. Where charts or photographs require to be made available to geographically dispersed individuals or companies, in order to meet close deadlines, there is a clear case for investing in some sort of facsimile provision. In such circumstance, the manager will need to consider such factors as security (who will be allowed to see the newly-generated items?) and whether use should be made of the PSTN or leased lines.

NETWORKS

Networks can be large (eg wide-area national or international) systems or local configurations for a factory or office complex. Local area networks (LANs) are seen as particularly relevant to the needs of the integrated electronic office. A LAN can feature as an element in the communications architecture of an office which includes several linked electronic workstations, and where the LAN itself is linked to an office controller, the Communications Management Computer (see Figure 2.1).

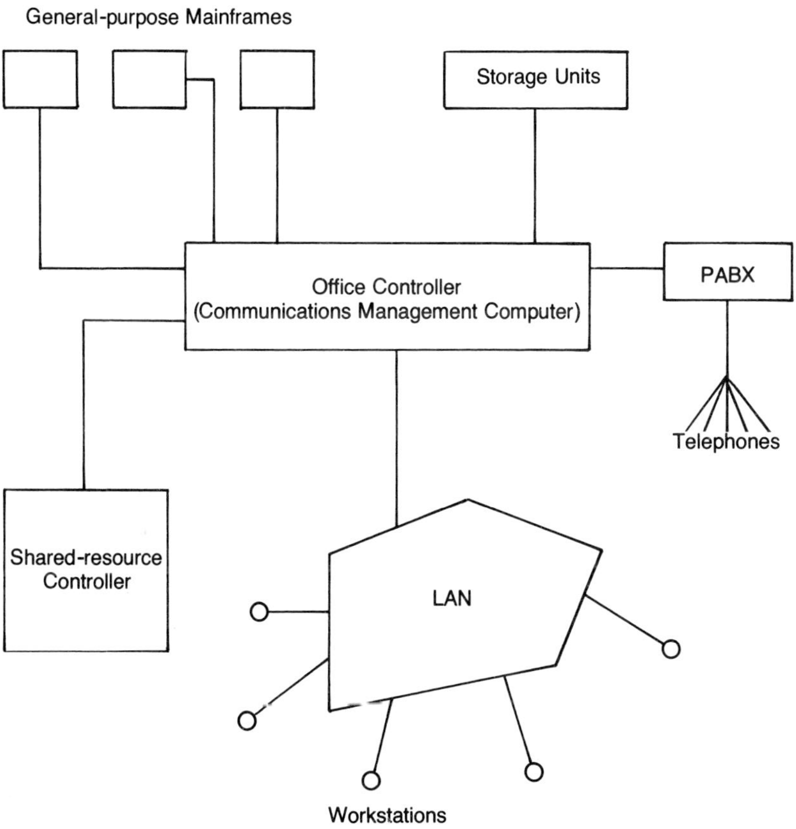

Figure 2.1 Communications Architecture for Electronic Office

Organisations are adopting various approaches to the development of internal communications architectures. Some are planning to implement a LAN approach or a PABX-controlled network approach (one option is that an Ethernet-type coaxial LAN be installed). The architecture shown in Figure 2.1 is a hybrid approach, using both PABX and LAN facilities. Here the shared resource controller can include such elements as: a facsimile machine, a graph plotter, a graphics printer, an OCR reader, image scanners and printers, disk and tape storage, and an intelligent photocopying machine. The local area network would typically connect to word processors, data processing terminals and multifunction workstations. (For more information about local area networks see, for example, Gee 1983 and also the Market Report, Number 65, January 1984, issued by Wharton Publishing.)

The multifunction workstations will typically provide facilities such as:

— document preparation using word processing software;

— document filing and retrieval;

— access to management information;

— various decision-support systems (eg Visicalc for manipulating tables of information, or a high-level programming facility such as APL);

— various administrative support systems (eg diary and 'tickler' facilities);

— electronic mail;

— a link with the Telex network;

— a link to the (internal and external) telephone systems;

— a voice-message system;

— a provision for voice annotation of documents.

The multifunction workstations – typically sited on (or near to) the desks of managers, professionals, personal assistants and secretaries – will be important units in the communications networks. They will largely define the scope of the more sophisticated electronic offices.

SUMMARY

This chapter has profiled some of the more obvious facilities that characterise the electronic office. There is a wide range of office functions that can be performed directly by, or with the aid of, computer-based systems. But these systems represent an extremely diverse spectrum of possibilities. Some of the facilities (Telex, word processing, facsimile, etc) have now been used for many years by thousands of companies and other organisations: experience has been built up and there is much available information about the scope of particular systems. Other facilities (electronic mail, teleconferencing, etc) are more recent innovations: less experience has accumulated and the scope of systems is still being explored in fundamental ways.

The disparity in the character and scope of individual office-automation systems is reflected in price, a crucial concern of managers charged with the task of developing a programme for the implementation of an electronic office for their organisation. Again we may emphasise that, in implementing office automation, managers should:

— gain an understanding of the full range of computer-based systems available for the electronic office;

— assess the needs of their company (or department) with care, allowing for future expansion (or other organisational changes);

— work towards the implementation of an integrated electronic office that aims to bring productivity benefits to all levels of staff, at the same time (hopefully) improving the state of staff morale.

It is useful to look in more detail at how the transition to an electronic office should be handled. Managers need a strategy for change.

3 A Strategy for Change

INTRODUCTION

The introduction of electronic systems into the office environment involves a number of steps or phases (these are described below). At every stage many factors need to be considered: the successful automation of office functions is a complex task. There may have been considerable investment in existing equipment, with limits on the extent to which it can be adapted to new working practices. Office procedures, formal and informal patterns of communication, and particular management styles – all of which may have evolved over many years – cannot be suddenly transformed without threatening the efficiency of both the department in question and the broader organisation. Electronic systems need to be evaluated; staff need to be trained; new management structures need to be developed. The manager needs to plan carefully for innovation in the office.

The broad aims of office automation are to improve efficiency and thereby to improve productivity. But there are circumstances in which efforts to automate an existing (inefficient) operation can simply produce greater inefficiencies. Conversely, excessive automation can be highly detrimental to an office that is already efficient, meeting the needs of customers, and operating with good staff morale. Every office is a dynamic, evolving system, shaped by a unique combination of factors. The uniqueness of the individual office must be fully recognised in any attempts to introduce or expand the provision of electronic systems. In particular, the task of innovation should not be seen as solely a technical matter. The manager should remember that the introduction of new technology has many human and organisational ramifications: every facet of a complex scene needs to be given careful attention.

A coherent management strategy for the introduction of office automation is essential. It must be clear exactly what the proposed innovation entails: no loose ends must be left 'to take care of themselves'. The scheme – in its various technical, staffing and organisational aspects – must be defined in detail. Otherwise money will be wasted: effort will be duplicated, there will be investment in incompatible systems, and (existing and newly-appointed) staff will be unable to operate the new systems successfully.

At the same time the detailed plan for innovation must allow for a flexible and creative adaptation to changing circumstances. Even a well-conceived plan will be unable to take into account *every* relevant factor: staff mobility, the rapid pace of technological change, the vagaries of the market – all introduce unpredictable elements which may not have been *covered* by an implementation plan but which should not *negate* its main terms and recommendations. Moreover a degree of flexibility in the plan leaves scope for individual initiative: the commitment of affected staff is an important ingredient in the successful implementation and operation of office automation (see also Chapters 7 to 9).

It is also important to realise that a management strategy for office automation is equally essential for both small and large organisations. The size of the organisation will necessarily influence the *character* of the strategy, but no organisation can be indifferent to misplaced effort, the acquisition of unsuitable equipment, the development of inappropriate management structures, and the waste of funds that these aberrations involve. A small organisation may have very limited investment potential – so an optimum distribution of funding is essential. A large organisation, with a greater capacity for investment, may introduce massive plans for innovation (in the office complex and in other parts of the organisation): such plans will inevitably have wide-ranging consequences for staffing, productivity, organisational change, etc. Hence any organisation, of whatever size, needs a coherent strategy for innovation.

THE STRATEGIC FRAMEWORK

The broad strategy for introducing electronic systems into the office environment can be considered to have six elements: diagnosis*, design, development, transition, implementation and adaptation (see Figure 3.1). (This follows the approach described in detail by Peltu, 1984.)

We may assume that the innovation strategy begins with diagnosis (see below, Diagnosis: Objectives and Targets) and then proceeds through the other phases (see below). But once the loop has been entered, via the first diagnosis, it is possible to move in either direction (hence the two-way arrows in Figure 3.1). The

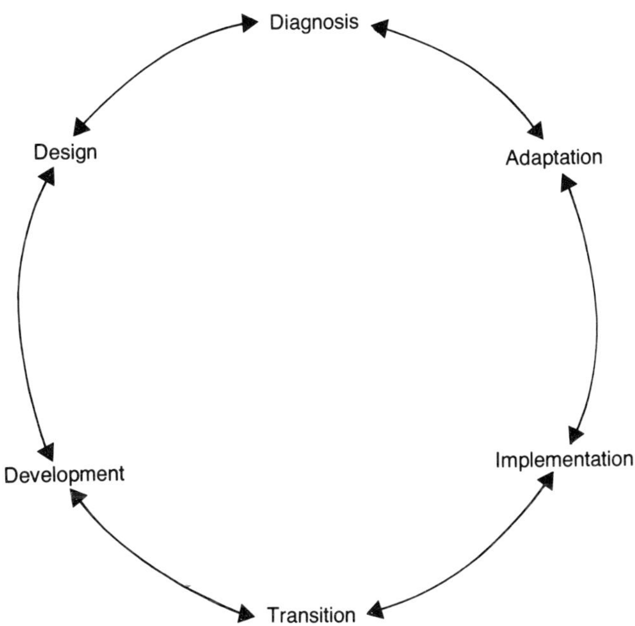

Figure 3.1 Basic Strategic Elements in Developing Office Systems

*The diagnosis phase can proceed by means of a feasibility study (see Chapter 5).

process of innovation in the electronic office should be regarded as an evolving, 'circular' activity, a process affected by many interactions, functions, pressures and requirements. The various stages in the process overlap and interact with each other. The management strategy needs to be continuously applied to the development of office systems. It is useful to consider the various elements in the strategy in more detail.

DIAGNOSIS: OBJECTIVES AND TARGETS

An important task in this phase is to define the purposes of introducing office automation. What are the aims and objectives? And how are these to be achieved? Answers to such questions must be provided within the broad strategy of the organisation.

The broad goals must relate to such questions as productivity and profitability: in general the organisation is aiming to survive, and possibly expand, in a commercial environment. But the broad goals are served by various subsidiary objectives (some of these may be achieved in parallel or a suitable 'chronology of targets' may be developed).

It is necessary, for instance, for staff and managers – albeit to different extents – to gain 'hands-on' experience of the new technologies. Suppliers, interested in possible sales, may help to facilitate this – either by running suitable demonstrations on their own premises or by bringing suitable equipment to the premises of the potential client. Thought should be directed at how the new technology can improve existing services and develop new ones. Can the new systems help to streamline the organisation, to solve particular problems, to improve staff morale, to improve management decision-making, and to stimulate fresh thinking within the organisation? It may be that the new computer-based systems will accelerate trends towards operational decentralisation, at the same time giving senior management more insight into what is happening in particular departments.

The diagnosis phase, providing an entry to the overall development loop (Figure 3.1), allows managers to stand back and consider where the organisation is heading (and where it *should* be heading). The systematic scrutiny that this entails can often lead to organisational improvements unrelated to the prospect of office automation: the diagnosis *may* yield the conclusion that automation is inappropriate in some areas of office activity and that desired objectives can be realised by improvements in existing systems.

The strategy should embrace the policy for the organisation as a whole, ie the corporate policy towards information management in general (this may also include telecommunications and data processing). And care should be taken to give attention to the long-term implications of the various technical, personal, organisational and economic factors. The aim should be an integrated system that meets the defined requirements. Mechanisms should be established to co-ordinate information resources and to lay down specific aims in particular areas: for example, standards will be needed to ensure that the various systems are technically compatible, though such standards may be waived in the case of systems to be used at a local level or for parochial personal reasons (care should be taken that any such exemptions do not give an excuse for failing to apply tight standards where they are necessary). The strategy should be concerned, at this stage, with broad objectives: more specific targets can be developed later.

It is also necessary to develop a management structure to implement the strategy. Mechanisms for co-ordination and control should be established, with appropriate managers given the necessary authority and resources. Senior management should be sensitive to the interpersonal tensions that may arise when managers are appointed or given fresh duties: leadership of an office automation project can be seen as an important power base. The person ultimately in charge of office automation is likely to be the managing director (in a small company) or a director (in a large company); and a top-level management group should keep a close watching brief.

There are advantages in keeping the management and development of office automation within the organisation: for example, in-house managers and staff can then develop the necessary knowledge of the new technology. Outside consultants can be used when there is a need for specialist skills on a short-term basis. Here, however, the consultancy background should first be researched; there should be agreement that an inadequate or unsuitable consultant can be replaced; and the task of the consultant should be carefully defined (there will be a requirement, for example, that the consultant convey all relevant information to in-house managers and staff). It is likely to be desirable to incorporate the consultant into the in-house management structure for the duration of the consultant's in-house activity.

The diagnosis phase is a 'broad-brush' phase in the overall strategy. It sets the scene for the various design and implementation phases that follow, and will be subject to various iterations (see Figure 3.1) as the market scene and the company potential change.

DESIGN: STAFFING AND COSTS

Before the strategic objectives can be translated into detailed system designs (see Chapter 5), ie before *feasibility studies** can be conducted, it is necessary to decide how the design will be undertaken. It is important to remember that the resulting office systems will involve more than items of the new computer-based technology. Attention must also be paid to operational procedures, organisational structures, staffing, job designs, and the physical arrangement of the working environment. This suggests that many groups in the organisation will be affected by the proposed office innovations, and should therefore be involved in the design of the new systems. Various ways of achieving this participation should be explored.

When computer-based systems have been implemented in the past, it has been common to rely excessively on technically-oriented systems analysts. Managers and staff have often felt left out of the consultation process, and have been encouraged to view the new systems as alien intrusions to which they had no real commitment. This meant that large numbers of staff had little interest in working towards the success of the new systems. To overcome this problem, it is useful to establish mechanisms that involve more people in the design, development, selection and implementation of systems. People likely to be affected by proposed changes should be invited to give their views and should be

*See Chapter 5.

kept informed of progress. Trades unions have proposed *technology agreements* requiring that staff be informed of possible technological changes before final decisions are taken. However, management itself should be encouraged to regard the commitment of involved staff as a prerequisite to the successful implementation and operation of new office systems. In this, management and unions may be seen as having many common interests.

The first task in the design phase is to analyse the existing modes of office working. For example, attention should be given to patterns of staff interaction, how costs are incurred, the types of operational difficulties that arise, and how services are provided. An effort should be made to quantify such factors as:

— the amount of material that is generated over a typical week or month (eg the number of words typed, documents produced, client records compiled, etc);

— the number of transactions handled (eg queries answered, orders processed, messages despatched, etc);

— the amount of information stored (on paper, microfilm, magnetic disk and tape, etc);

— the number of staff at various levels;

— the costs of staff, telecommunications facilities and other elements.

Analysis of existing working patterns should also include scrutiny of organisational structures, formal management reporting routes, and standard working procedures. And managers and system designers should also be aware of informal methods of communication and working practices: chance meetings, friendships, personal ways of adapting formal rules to local needs – all are important to the working of the typical office but are likely to elude the neat theoretical frameworks constructed by pedantic analysts.

Attention should also be given to the quality of service provided, the character of management decision-making, and whether the working environment is conducive to successful operations. Again, care should be taken to examine both formal and informal pressures and interactions: difficulty in quantifying a factor does not mean that it is unimportant. A system may well be unsuccessful if too much emphasis is placed on formal and quantifiable factors, ie if elements of the real world are ignored.

Costs and benefits of the new technology should be estimated on a realistic basis. For example, one set of resources may be replaced by a cheaper set: jobs may be abolished, with the productivity of surviving staff improved (here there may be savings on accommodation and other overheads but unquantifiable costs in terms of morale). Efforts should also be made to identify new products and services, new job possibilities, etc. Managers should be sensitive to unexpected potential that may emerge in new electronic systems, and equally sensitive to the costs that may accrue if staff do not feel encouraged to commit themselves to change.

DEVELOPMENT: SYSTEMS AND SUPPLIERS

Managers should be able to handle the various aspects – technological, human and organisational – of systems development. They need not be technologists themselves but should know enough to be able to assess, possibly with in-house or independent assistance, the competing claims made by outside technologists and suppliers. To control the development process, managers should:

- establish the criteria to be used in the various phases of development, implementation and operation;
- hold frequent progress meetings with development staff;
- insist that the technologists explain problems and progress in non-technical language;
- insist that any design changes be systematically reported.

We may expect systems development to be less onerous in a small organisation than a large one. A transition to office automation may involve simply acquiring a word processor for a small office; or it may involve massive and far-reaching change in organisational structures, job definitions and operational procedures.

System suppliers should only be sought after a clear specification of user requirements has been formulated. The *requirements specification* should cover such elements as performance needs, required system functions, necessary maintenance provisions, details of available finance, technical standards to be met, and the possibility of future system enhancement or expansion. The specification can then form the basis of an *invitation to tender* which can be sent out to selected potential suppliers; and can also form the basis of the contract agreed with the selected supplier. (Choosing office systems is described in more detail in Chapter 6.)

Both systems and suppliers should be assessed with care. There are many systems on the market and suppliers vary enormously in what they offer. Do they, for instance, offer a maintenance service, equipment replacement provisions in the event of catastrophic breakdown, training for the customer's own staff, advice on possible system expansion routes (possibly involving recourse to equipment from other suppliers)? It should also be remembered that today there are high-quality systems available at relatively low price. High cost does not guarantee good quality or inevitable suitability for the user's department or company; low cost does not necessarily mean that a system is inadequate for envisaged applications.

TRANSITION: MANAGING CHANGE

It can be difficult to transfer from old to new systems. For a time some applications may be running on the old system, others on the new. Staff will become confused and discouraged if clear guidelines for the transition are not laid down – and it is very important that staff discouragement be avoided: jaundiced attitudes at this time will be hard to eradicate and will affect the whole future operation and expansion of the system.

Again managers should be able to operate through both formal and informal channels to ensure that the transition is conducted as painlessly as possible: there should be frequent communication with staff to ensure that progress is satisfactory, and there should be receptivity to staff suggestions and observations. Staff should be encouraged to voice doubts and anxieties; otherwise, at best, resentment may fester or, at worst, there may be high rates of staff absenteeism and staff mobility or even malicious sabotage! Managers should also be aware of possible staff rivalries during the period of transition, a circumstance that can easily cause recalcitrance and a reluctance to co-operate for the overall benefit of the department or company.

The transition is likely to involve the translation of paper-based information into a computerised form, or data may be conveyed from one electronic system to another. This can involve adapting or rewriting software, translating information from one computerised standard to another, retraining some individuals and recruiting others. There may be a need for new office furniture and new heating and air-conditioning systems, even alterations to the fabric of the building. Transition, a stressful time for all concerned, should be planned for in a careful way and implemented in a systematic, step-by-step fashion.

IMPLEMENTATION: THE WORKING SYSTEM

The acquisition of any electronic equipment for the office demands careful consideration, and this is particularly so where computer-based facilities are to be brought in as part of a scheme for office automation. Even the introduction of a microcomputer or a small business system demands careful thought about the likely impact on existing services and working practices. A relatively small innovation may represent the first step in a progressive programme of automation.

Again we may stress that transition from the old to the new system can be a tricky task. Data must be transferred; relevant staff must be trained; and other affected personnel must be told about the likely impact on their activities.

Before the system is accepted from the supplier, it must be tested against the criteria defined in the design specification and included in the contract. The system should be regularly monitored to ensure that all aspects of its performance meet the design requirements.

The system may seem well suited to the company's needs when first introduced, but there may be unexpected consequences when, for example, a key member of staff resigns, efforts are made to broaden the company's product range, or there is a sudden economic shift in the marketplace. Moreover, enthusiasm with a new system can wane. Staff may become complacent and allow inefficient practices to creep in. A mechanism for revealing possible inefficiencies quickly will allow corrective action to limit the damage to the services provided by the department or the company. Some Japanese firms have adopted a 'quality circle' approach to efficiency: all staff participate in the circles, offering practical suggestions on how the efficiency of operations can be improved.

There is also the question of system security. We have seen (Chapter 1) that information is properly regarded as an important economic resource. Many computer-based systems contain information that is vital to the company, customers, clients, employees and other organisations and individuals. The operation of the organisation and the well-being of personnel could be seriously threatened if the computer-based systems are adversely affected by natural disaster, accident, criminal behaviour or other forms of malicious activity. (NCC has published more than a dozen books on computer security: see, for example, Michael Wood's *Introducing Computer Security,* 1982.)

The system design should include provisions for adequate security (see, for example, Wood and Elbra, 1984). Attention should be given to such aspects as the provision of duplicate back-up equipment for key parts of the system, setting performance reliability criteria, writing security precautions into the software, and establishing good physical security precautions. The security provisions should be continuously monitored: complacency and carelessness in such matters can be very costly. The ability to trace a clear audit trail through the system is a useful contribution to security.

ADAPTATION: LEARNING FROM EXPERIENCE

However sound the initial system design, there will be an inevitable requirement that practices and provisions adapt in the light of circumstances: there will be a need for individuals to learn from experience, and to take appropriate action.

Much has been said (and written) about how individuals and organisations move up a 'learning curve' with computer-based and other types of systems. When a system is first installed, it may begin operations in a hesitant, uncertain fashion. Operators, though well-trained and knowing what to expect, will soon learn that theoretical expositions do not always accurately or fully depict the contingencies of the real world. It is inevitable that in the early stages of system operation mistakes will be made and the system will fail to realise its full potential. Soon, however, learning speeds up, fewer mistakes are made, and the system begins to provide its full benefits. One sign that the system is 'coming into its own' could be that it is now being addressed to the new services, new jobs, new working methods, etc, that were envisaged: the system is no longer being used simply to mimic the old, obsolete ways of working.

Problems should be openly discussed and analysed. One aim should be to encourage imaginative applications of the system, possibly transcending the system uses originally envisaged in the diagnosis phase. In this way a progressive entrepreneurial spirit can be cultivated: the involved staff can be encouraged to make the new system work for the success of the department and the organisation. And managers should be aware that changing circumstances can stimulate a further iteration of the strategic circle (Figure 3.1), starting with the diagnostic phase.

SUMMARY

A strategic framework has been identified, consisting of six overlapping phases:

— diagnosis;
— design;
— development;
— transition;
— implementation;
— adaptation.

These phases must be worked through in a careful, step-by-step fashion. Action should not be taken without due thought, but the need to plan should not be allowed to serve as an excuse for tardy action. The long-term strategic aims should give priority to user requirements: managers should remember that the office exists to provide a service.

It should also be remembered that the new system is more than simply a configuration of technological equipment. The technologists should be kept under firm management control, and should, in particular, communicate with managers in non-technical language. Managers should establish mechanisms for monitoring progress, should regard learning-from-experience as extremely important to the success of the system, and should revise the strategy as required. And all costs, not just the more obvious ones, should be considered: the main overriding reason for introducing the system is to improve the profitability of the organisation.

Part 2
Towards Office Systems

4 The Office in Transition

INTRODUCTION

New office systems should not be introduced solely to automate existing ways of working. The new computer-based facilities can stimulate new modes of operation in the office, bringing fresh benefits and exploiting the creative initiatives of involved personnel. The new technology demands a revision in office routines, management procedures, and organisational structures.

This chapter discusses management and organisational features of the office in transition. Hierarchical structures are traditionally common in companies and other organisations, and it is important to realise that the new office technology is likely to affect every level in the hierarchy. Office-system innovations may render traditional organisational boundaries inappropriate and may demand new management structures.

Attention is also given to aspects of organisational procedures (these can be programmed into computer-based systems and progress can be monitored). It is recognised that work organisation and performance will change, and this will influence how work units are established and how they will co-operate. New organisational methods will be needed to manage the information sources efficiently. And it should also be acknowledged that the innovation process will have a 'political' dimension: managers should have the authority to handle the technical questions and also the intra-organisational power battles that may arise.

This chapter also emphasies that office innovation is bound to affect the relationship of the department with customers, clients, suppliers and other parts of the organisation. It is necessary to consider how the service to the customer will be affected by any proposed introduction of electronic office systems.

SERVING THE CUSTOMER

The introduction of new office systems will only be successful if the services provided by the office are improved. For example, does the new technology make it easier for the organisation to deal with customers? Do the customers themselves perceive an improvement in the services they are offered? Are there now manifest advantages over the old system? Can suppliers now deal with the organisation in a more effective way? And do other departments within the organisation now enjoy improved services?

It should be recognised that both staff and customer are likely to be disturbed by any transition to new working methods. Existing methods may have been in operation for many years, and individuals may feel uneasy about any radical innovations. For this reason, care should be taken to emphasise the benefits of the new systems: the new facilities should be 'sold' with enthusiasm to both in-house staff and external personnel who may be affected by the new developments.

A key objective is to improve the office services offered to the outside world. Management should develop a plan to ensure that this objective is achieved. Attention should be directed at:

— how current services are provided by the office, how such services can be improved, and how the new technology can contribute;

— how the new systems can make it easy and appealing for computer-naive external people to interact (using, for example, a cash dispenser or a visual display unit) with the new technology;

— the provision of training for in-house staff who come into contact with the outside world. The purpose should be to enable staff to promote the advantages of the new procedures, to handle problems promptly and diplomatically, and to be able to take appropriate action (repair or providing an alternative service) if there is a failure in the system;

— the development of an awareness programme to convey to outside clients, customers, suppliers, etc the reasons for change, the benefits that the new system will bring, and details of new operational procedures that may affect how external users exploit the available services;

— the provision of suitable promotional material at the point of service;

— the development of reliable and effective interfaces to the outside world. These should be systematically monitored to ensure that they are meeting the requirements. It is useful, for example, for managers or other in-house staff to take the role of a customer or client to experience directly the quality of service being offered to the external world.

There are many possible users of the new systems, and their individual requirements should be taken into account. Some customers or clients ('one-time') will only use the services on a single occasion (for example, to buy an insurance policy or to seek advice from a solicitor). Others are regular but infrequent customers or suppliers (those booking holidays, visiting a doctor or supplying special goods). Regular and frequent customers and clients include customers in a bank or building society, recipients of unemployment benefits or pensions, and suppliers of regular goods.

The needs of in-house staff should also be considered with care. Other departments, branches and projects within the organisation will expect to use the office facilities, and individuals will want access to their own payroll details, personnel records and other private information that relates to the terms of their employment.

MANAGING THE ORGANISATION

It is important to establish the procedures and policies that define how the organisation should operate and where the responsibilities lie in the circumstances of the new technology. Attention should be given to, for example, the management structure, informal factors, and ways of organising specific office tasks.

The Management Structure

A particular management style or philosophy can be discerned in any organisation (see also Chapter 7). This style derives from management beliefs and attitudes that are conditioned by many factors, only one of which is the impact of the new technology. There is, for example, the inevitable tension between the need of senior management to maintain central control and the need of middle managers for local autonomy in decision-making. This tension, handled in different ways in different organisations, is affected by the facilities offered by the new technology. It is significant, for instance, that the large computers of the 1960s were highly suitable for centralised management structures, but that modern systems provide flexible options (for example, via distributed processing) for any management structure.

The new technology is one powerful factor encouraging a re-evaluation of existing management structures. Traditional organisational hierarchies have allocated responsibilities unambiguously, with each person reporting to one manager (the chief executive may report to a board). Within the hierarchy a duplication of information may occur, different departments requiring the same information for different purposes. Communication tends to be vertical in the hierarchy, rather than horizontal. One characteristic of the new office technology is that it can cut across vertical structures.

It is conventional for an electronic data base to store information centrally but to make it available to local units as necessary: for example, manufacturing, sales and administrative departments can be directly linked, a possibility that encourages a revision of traditional control and co-ordination methods. In some instances, it may be convenient to allow a person to report to two or more managers for different tasks. This form of *matrix* (or *network*) management encourages interaction between units, a circumstance that can be helpful during a period of organisational transition.

Managers considering organisational structures should analyse how the technology can influence such elements as the boundaries between functional units, the interactions between units, and the operational scope of individual units. The traditional responsibilities of staff, managers and units may need to change so that new opportunities can be exploited to advantage. In such circumstances, it will be necessary to scrutinise the quantity and quality of staff required for each unit and activity, and to acquire – either through retraining or recruitment – new managerial and operational skills. The immediate impact of the new technology on management structures should be analysed in detail, and estimates should be made of what the consequences will be for the future. New management structures should be flexible enough to allow for change.

Control within the Organisation

Organisations cannot operate effectively without a framework of rules, procedures, reporting mechanisms, terms of employment, etc. Such devices allow information to be gathered on performance (of individuals, departments, the company as a whole) which allows problems to be identified and rewards (and reprimands) to be given. A wide range of committees, meetings, projects, etc use the various devices to manage the many activities taking place within the organisation. The new technology can affect this situation in many ways.

Formal routines and procedures can be built into computer software to provide *preprogrammed control:* users of the computer-based systems are then required to follow the procedures, with any deviations automatically monitored and reported. The systems can also provide management with *full and speedy information* on performance, allowing 'fine tuning' controls to adjust the organisation to a rapidly changing world. *Interactions between groups and individuals* can be made more numerous and more complex by means of such electronic-office facilities as information networks, data bases, facsimile, electronic mail, teleconferencing, etc. Communication can thus take place across geographical and organisational boundaries in ways that would have been impossible with earlier systems. *New groups* (for example, to include people with a common professional interest) can be formed.

The various organisational ramifications of the new technology will have different consequences in different companies. Some management methods will continue as before, not 'by default' but following a close scrutiny of the available options. Managers should be aware of how the new electronic systems can affect traditional methods of control and co-ordination.

The Role of Middle Managers

In small organisations, top managers tend to communicate directly with a large proportion of the staff: the managing director of a small family business may claim to know all the employees by name. In larger organisations, 'middle managers' serve as intermediaries between top managers and the staff at the operational level who are carrying out the required day-to-day tasks.

The middle managers – sometimes decried as part of a 'middle manager bulge' in the company – are concerned with filtering information, decisions and working requirements from senior management down to the operational levels; and with reporting information, results and requests from the operational level up to senior management. A key feature of the new-technology situation is that, for the first time, senior managers in large organisations can be put in direct communication with the operational levels. Senior managers can now exploit decision support aids to interpret up-to-date performance and trend information. This means that senior managers are being helped by modern technology to monitor closely every aspect of the organisation's activity, with the corollary that there is a diminished need for some middle manager resources. The ease of communication between top and bottom which formerly only characterised small organisations is now possible, via new technology, in large organisations. This possibility will inevitably affect the shape of companies and other organisations in the future.

Organisations may come to abandon the bulge around the middle, and loosely structured arrangements may emerge where shifting relationships between units, allowing for flexibility and adaptation, may be regarded as advantageous. In any event, the broad requirements in the successful organisation still need to be met. Information and other economic resources should be exploited in a profitable way, but one that is not counterproductive in the long term (ie there should be due attention paid to expansion potential, adaptive flexibility, and the morale of surviving staff).

Informal Factors

There is always a discrepancy between a *formal* description of how an organisation works and a description that takes *informal* factors into account. A formal description (sometimes a pious hope) relies upon such elements as recorded corporate procedures, administration manuals, organisational charts and job descriptions. A fuller, more accurate, description, while not ignoring such elements, gives attention to a wide range of factors that may not be easily quantified. It is important that the benefits of the many informal factors are not lost when the new technology is introduced.

For example, informal methods usually develop because formal procedures are unsuited to particular circumstances (eg the changing relationships between the office and the users of its services, the specific working needs of individuals). This implies that informal methods can serve as an *adaptive mechanism,* helping the organisation to function efficiently when formal procedures lack scope or flexibility.

The ability to break away from formal procedures gives local management and staff an element of *autonomy* which can serve as an important motivational stimulus. Individuals are able to take creative initiatives designed, in part, to keep the organisation in touch with local needs. Staff are also free to benefit from *chance encounters* that occur during lunch, at social gatherings, or in other circumstances not thought immediately relevant to practical operational activity.

Informal discussions, 'off-the-record', can generate useful commentary on proposed ideas and formal programmes. Here individuals may be less inhibited than when asked to contribute at a more formal meeting. Information may emerge during an informal discussion which may not be elicited in a formal framework.

There is a danger that the introduction of electronic-office systems could destroy some of the benefits that derive from informal ways of working. It may be tempting for some managers, when planning for office automation, to develop uniform, inflexible, preprogrammed methods of working and organisational structures. For instance, when all communication is expected to occur via fully-monitored formal communications channels, various informal modes of interaction in the office environment will be inhibited. This could cause a variety of creative benefits to be lost to the organisation.

In systems design and development, attention should be paid to both formal and informal factors. An effort should be made to evaluate the benefits that have traditionally derived from the operation of the various informal mechanisms. When such benefits are more clearly appreciated, it is easier to preserve them when new office-automation systems are introduced.

Organising Office Work

Traditional office work has been relatively unstructured (compared, for example, with the unswerving attention to routine expected of production workers). When office automation was first introduced, the focus was on applications that lent themselves to formal design and specification techniques. In applications such as accounts and payroll, it is possible to structure files, categories and formats unambiguously. Many tasks, however, that confront the modern office have unpredictable features that may defeat a nicely formalised approach: there may be unexpected queries or problems to be handled. This has allowed the best managers to retain flexibility in handling new situations, at the same time remaining willing to adopt a formal structured approach where this is appropriate. Office technology should be seen as aiding both structured and unstructured activity.

It is important to decide, in the circumstances of the new technology, how to organise the activities of secretarial, clerical and other support staff. Early efforts at office automation focused on simple, possibly simplistic, productivity criteria. For example, the productivity of typists was seen as relevant to assessing the success of word processor installations. It was thought that word processors should be established in a central pool, with the number of characters typed per operator regarded as a critical measurement parameter. However, when attention was paid to broader criteria, it was seen that various disadvantages derived from a centralised

WP facility: the effectiveness of managers and staff was diminished, and overall efficiency was reduced. The close personal interaction between people had been damaged, with – as one consequence – senior managements being forced to spend their time photocopying, filing and carrying out other important but trivial tasks. At the same time, isolated from the professional staff, the WP operators lost interest in what they were typing. Typing errors increased, and an excessive number of document drafts were requested.

An alternative to the old centralised approach is to establish an *administrative support* (or *executive support*) group. Here a number of secretarial, typing and clerical staff are brought together to share word processors and other new-technology systems. The members of such a group do not become isolated from the managers and other professionals whom they serve. Staff can be offered more varied work and a good level of morale can be maintained. In some cases a centralised pool may offer economic advantages, but it is also important that support staff remain close to managers. The individual organisation must decide what is the best arrangement for its needs – when *all* the relevant factors have been considered. Managers should be encouraged to take a long-term view: tempting, but simplistic, short-term criteria should not, in isolation, be allowed to determine a development strategy for office systems.

OFFICE INFORMATION SERVICES

The new technology facilitates the introduction of a wide range of unprecedented office services. These need to be co-ordinated within the context of general corporate requirements. For this purpose it is necessary to develop a co-ordinated approach for the management of all information resources, to decide how to manage the office technology project, and to provide the appropriate person or group with the authority and resources to implement the office-automation policies.

Information Management

Office technology can have an impact on many resources that have traditionally been handled by separate departments and on separate budgets. These resources include: office administration, data processing, telephone and telecommunications, mail, accommodation, and travel. There is a requirement, with the advent of the new technology, that these various resources be co-ordinated from the highest level in the organisation. Otherwise there will be an expensive duplication and waste of effort and investment. For instance, it is quite likely that different departments will develop incompatible subsystems leading, at some later date, to chaos and confusion. In fact many personal computers have been introduced into organisations on an ad-hoc basis, limiting the co-operative scope of such systems.

One approach is to establish *information centres,* 'open-house' units that are close to users of computer-based systems. Managers and other staff can approach such centres for advice and also obtain hands-on experience of computer systems.

It is important that departments and managers co-operate to achieve co-ordination of information management. Furthermore the impact of particular office systems on particular functions and activities should be noted. For example, information networks and teleconferencing can be used to cut travel expenses; the use of electronic mail can reduce the costs of the internal and external post; and the use of word processors can cut the typing load, and so reduce costs. At the same time, managers should realise that office technology needs to be compatible with existing DP systems; DP systems may sometimes have to be adapted to fit in with the office-automation programme; and important costs (for example, those incurred in making changes to the physical environment of the office) should not be ignored.

It may be necessary, to ensure adequate degrees of co-ordination, for the chief executive to assume responsibility for information management. A steering committee, set up at the highest level, may be necessary to co-ordinate the implementation and operation of the new office technology. And an information manager, possibly heading an information management group, may be required in order to report to top management. The mode adopted by a company will depend upon many factors (eg levels of in-house expertise, the size of the organisation, whether there are plans for rapid corporate expansion).

The Office Technology Project

The office technology project needs to be established within overall information management co-ordination. The project needs clear directives and competent leadership with adequate authority and resources. It may be necessary to establish temporary groups to see the project through to completion. Such groups would work, possibly on a long-term basis, with permanent departments until the project had achieved its objectives.

In recent years the responsibility for office-technology innovation has been variously assigned to:

- data processing staff and others with a computing background;
- managers with office-administration or O & M backgrounds;
- executives with general management competence;
- units or managers from outside the organisation.

Responsibility for office automation may be keenly sought by ambitious individuals who see this area as an important management power base. Senior managers planning the framework for an office technology project should be aware of these 'political' considerations. It may be difficult to decide who should head such a project.

Traditional data processing experience has usually evolved outside the mainstream of corporate activity. On the other hand, corporate executives may be daunted by the 'technological revolution' and may lack the sort of imagination that can envisage the possibilities offered by the rapidly evolving electronic systems. Ideally the individual or unit responsible for the office technology project should have *multidisciplinary skills* (to handle a broad range of technical, social, economic and organisational problems), should be capable of *imaginative and innovative leadership,* should appreciate the precise *corporate objectives,* and should have appropriate *authority and resources.*

The success of the office technology project is critical to the long-term success of the company. In the modern commercial world it is essential that firms learn to exploit the potential of automated office systems. A well-conceived and well-executed office technology project is essential to this purpose.

SUMMARY

This chapter has focused on a range of management and organisational matters. In this context, a number of guidelines have emerged; for example:

- informal, as well as formal, methods of working are important to the success of an organisation. Care should be taken, when introducing office automation, not to lose the benefits that derive from informal ways of working;
- the new technology makes possible a wide range of new services. It is important that these be introduced in a coordinated way. Individual departments should not introduce new office systems in a piecemeal fashion, with the duplication and waste of resources that this involves;
- everyone in the organisation is affected by the new office automation. Appropriate organisational structures should be developed with the working needs of all staff in mind. A comprehensive organisational approach – considering the roles of all staff, from clerks to senior managers – is essential;
- the new services and provisions should be 'sold' to all affected individuals, both inside and outside the organisation;
- an effective office technology project, well-conceived and well-directed, is essential to the successful implementation and operation of integrated office systems within the organisation.

The choice of office systems (Chapter 6) is an essential activity in the office technology project. Other key activities relate to the handling of human factors (see Part 3).

5 Approaching Systems Design

INTRODUCTION

The successful implementation of office automation involves, at some stage, paying considerable attention to the question of systems design. This is a complex task: the word 'system' can denote many different concepts and entities, all of which must be appreciated if office systems are to be introduced in a sensible way.

The overall office automation system includes machines, various means of connecting them to each other and to power suppliers, further devices to facilitate communication with internal and external users, the various elements of the office environment (building, ducts for cables, air-conditioning facilities, non-electronic furniture, etc), and people. System design must take account of these various elements.

Within the overall system are many configurations which are called *systems* in their own right. Thus there are mailbox systems, teleconferencing systems, facsimile systems, etc – and these also may be broken down into further systems (or *sub*systems). An electronics engineer or logic designer will be interested in the *hardware* system, the disposition of electronic components, switches, wires, screens, etc. A programmer will be interested in the *software* system, the range of coded instructions written to enable the hardware to function in various ways.

Hence, effective design for the office environment must take into account a wide range of different factors: equipment, power supplies, buildings, furniture, consumable materials, programs, interfaces with the outside world, job designs, people, etc. The *feasibility study* is the key initial step that aims to analyse the existing situation and define requirements before detailed systems design begins.

THE FEASIBILITY STUDY

The definition of a computer system should be seen as beginning with a feasibility study* which scrutinises all aspects of the situation being investigated. It should not be assumed, however, that a feasibility study will *necessarily* lead to the design and implementation of a computer system. The study may indicate that the departmental or organisational requirements can be met by a re-organisation or re-allocation of resources. Though it should be said that by the time the feasibility study has been authorised, the management may be expecting to invest in computer-based systems.

Managing the Feasibility Study

A single person, possibly an outside consultant, may be asked to carry out the feasibility study; or an in-house team, comprising a manager and staff, may be set up for this purpose. Ultimate responsibility for the feasibility study should rest with the managing director (in a small company) or a group director (in a large company). A computer professional should be involved, either as a team member or as a consultant, in the study. Clear objectives should be set by management, and the results of the study should be presented in a lucid report. Three principal types of feasibility evaluations should be made:

— *economic* (value-added benefits, cost-saving, improved productivity, equipment costs, personnel costs, etc);

— *technical* (systems strategy, information needs, available products, conversion requirements, relevant commercial services, etc);

— *social* (human factors in systems design, consultation with staff, new skills needed, required training, impact on working methods and organisational structure, etc).

*The feasibility study is the principal tool to be employed in the diagnosis phase of the strategic study (see Chapter 3).

Terms of Reference

A consultant appointed to carry out a feasibility study will be required to complete his task by a specified date. Similarly an in-house feasibility team should be regarded as a task force with a limited life. In these circumstances, an efficiently performed study requires that terms of reference be clearly defined. These should include:

- the scope of the study (for example, the activities to be investigated);
- the corporate and department objectives (where is the company heading?);
- relevant external factors (legislation, competitor activity, etc);
- involved staff (manager, in-house staff, external consultants, etc);
- performance criteria to be satisfied by final system;
- budgetary constraints on any recommended system;
- the resources and expected working methods of the study team;
- schedules, including 'milestones', on the route to study completion;
- format for interim and final reports;
- criteria and methods for evaluating recommendations;
- necessary procedure if terms of reference change during the study.

The terms of reference are established to try to ensure that the study team (or the individual consultant) focus effectively on the study requirements so that useful conclusions are reached within the time available. In short, the terms of reference aim to impose discipline on the feasibility study.

Topics to be Studied

It is necessary to decide applications to be studied and to determine the eventual schedule for computerising the activities (if the decision is taken to proceed). In this context three major aspects should be considered:

- the corporate requirements;
- understanding of particular applications or tasks;
- awareness of the potential of information technology.

The main applications to be studied in the context of the electronic office will usually be obvious (eg word processing, mailbox, facsimile, conferencing, etc). The study team may produce practical applications that were not envisaged at the start of the study. These may well be accommodated within the (suitably flexible) terms of reference, and may lead to further cost justifications for investment in computer-based systems.

Presenting the Study Conclusions

The conclusions of the feasibility study should be presented in a lucid form for evaluation by management and staff. Different reports should be written for the various concerned people (eg for the managers who will use the system, operators, the manager responsible for the study, senior managers, etc). The conclusions of the study report should relate directly to the original terms of reference, and should indicate the methods used, the people consulted, and the extent to which the proposed system falls within the budgetary constraints and meets the performance criteria. A typical report will include:

- an overall summary of conclusions and recommendations;
- the scope of the study, indicating whether the original terms of reference were fulfilled;
- the nature of the existing system, and which of its features should be retained and which modified or discarded;
- outline proposal for the new system, indicating how it will bring new benefits (cost savings, improved productivity, etc);
- equipment recommendations;

— personnel and training recommendations;

— schedules for designing and implementing the recommendations, and details of costs and resources required;

— guidelines on how external events may affect the implementation of the recommended system, and how system flexibility will allow adaptation to such changes.

At this stage, senior management is required to evaluate the recommendations and to decide whether to proceed to systems design and implementation. We would expect cost and profitability considerations, on both a short- and a long-term basis, to be key influences in this area of decision-making (cost elements of computer systems are shown in Appendix 5).

ELEMENTS IN SYSTEMS DESIGN

We have seen that there are many aspects to systems design (not least because 'system' can denote many different types of entity and configuration). Figure 5.1, showing the various tasks from the 'decision to proceed' to 'reviewing system performance', indicates some of the elements that need to be considered in designing the overall system. These include:

— design of the computer systems (attention to performance capabilities, storage capacities, available software, etc);

— preparing the computer site (provision of environment to house the computer-based systems, attention to equipment layout, air-conditioning, non-computer furniture, etc);

— organisation of the required staffing provisions (assignment of management, supervisory and other responsibilities, retraining, recruitment, etc).

The detailed design of computer equipment is essentially a matter for the manufacturer and supplier. The manager contemplating the introduction of office automation needs to have some insight into, for example, the performance differences between one computer or printer and another; but the manager need not become involved in the minutiae of electronic or logic design. The aim is to produce a requirements specification, following the find-

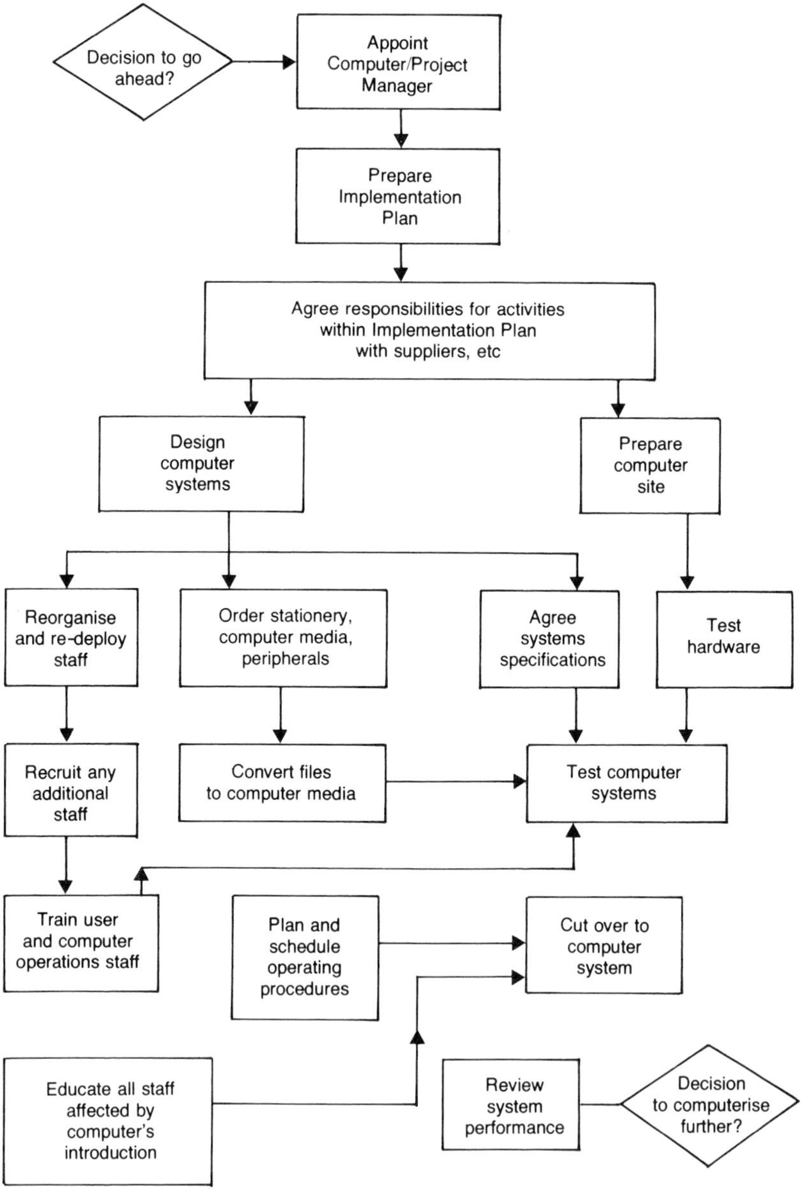

Figure 5.1 Stages in Design and Implementation

ings of the feasibility-study team, and then to invite appropriate suppliers to tender (see Chapter 6). At the same time it is useful for the manager to appreciate some of the computer design considerations that affect how particular systems are received in the office environment (some of these aspects are discussed below). Aspects of workplace design are also considered in the present chapter.

The organisation of the necessary staffing is essential to the success of the electronic office. Expensive equipment will at best not realise its potential and at worst lie idle if appropriate attention to the human factors in the situation is not given. The design of the overall office system, automated or not, demands attention to the roles and morale of the staff. The commitment of personnel at all levels is crucial to the success of any scheme for introducing computer-based systems into the office environment (this aspect is discussed in more detail in Part 3).

Design for the User

To design the complete office system involves taking a number of specific design decisions, some of which relate to the roles and responses of the user in the working situation. It is necessary, for example, to design the jobs that people will perform in the automated office. The aim of *job design* is to define the tasks that are the allocated responsibilities of specified individuals. Even a good *technical* system that adversely affects jobs will run into difficulties. User apathy and hostility will lead to staff mobility, absenteeism, sickness and even sabotage. The system may be used in unplanned ways or some planned facilities may be ignored.

Job designers should aim to provide staff with a degree of autonomy that is commensurate with responsibilities and abilities. Ideally there should be variety, mental stimulation, and the opportunity for learning. The work should be meaningful, and the individual's efforts should be recognised and rewarded, as a means of maintaining a good level of morale.

There are various practical ways of organising tasks into jobs. For example, managers may introduce job rotation, job enlargement or job enrichment (which speak for themselves). Use may be made of autonomous work groups, assumed to have more scope and creative potential than a single individual. And work may be divided by function, by supplier, or by product type.

It may also be considered useful to organise mechanisms for *activity feedback* at various levels: people are likely to perform better if they are provided with feedback on their performance, and generally kept in touch. Thought should be given to the required levels of activity feedback and to the design of activity feedback information.

In addition to the various aspects of job design, it is necessary to consider *dialogue design*. This specifies the content and format of communication between users and computer systems. With many systems in the electronic office, dialogue design focuses on the use of keyboards and visual display units (VDUs).

The dialogue designer needs to know details of the user activities, the roles of dialogue users, jargon (and abbreviations, codes, etc) understandable to the user, details of the equipment, and details of existing dialogue design standards. It is important to allow sufficient time and effort for dialogue design. The existing standards (eg position of transaction codes on the screen, signing on/off procedures, error messages, use of blinking or reverse video, etc) should be exploited and adapted as required. And dialogues should be flexible (for a fluid population) and logical (for ease-of-use). (Dialogue design is described in detail in Damodaran et al, 1980, Chapter 4.)

It is also necessary to design procedures for faults (which reduce the efficiency of systems) and breakdowns (which render systems inoperable). *Procedure design* in this context should aim to allow the user:

— to deal with the problem (rectifying the fault or breakdown);

— to take appropriate action to safeguard the business operation;

— to feel in control of the system;

— to record the problem for long-term statistical analysis.

These various design aspects – focusing on jobs, dialogues and procedures – are crucially relevant to how the staff perform in the working office environment. They determine thereby, along with a wide range of other design tasks, the extent to which the new office-automation system is successful.

Workplace Design

The various features of the workplace should be designed as part of the overall system design. For example, attention should be given to how individual workstations should be laid out and positioned; and thought should also be given to such aspects as lighting, the room climate, noise, and the room layout. A preliminary survey should collect data on the dimensions of the workplace, the layout of furniture, access requirements, floor characteristics, and the position of air-conditioning and heating equipment.

Workstations should be designed to take into account a wide range of ergonomic (or human-factors) considerations. Care should be taken, for instance, to avoid workstation designs that cause bad body postures, poor visual environments, and other health or safety hazards (see also Chapter 9). The design of workstations has been divided into six stages:

— establishing initial design parameters;

— establishing worksurface dimensions and layout;

— designing the chair and footrest;

— designing the overall furniture layout;

— checking the design;

— testing the design in the working environment.

Design should take place using anthropometrical and psychological data for a specified user population (eg male and female adults, below the age of 65 and having no physical handicap). During the design process, incompatible requirements may be encountered, and optimum compromises will need to be made (workstation design is discussed in detail in Damodaran et al, 1980, Chapter 5).

Where artificial lighting is required it should be sufficient for the tasks being performed, but should not cause glare or induce visual fatigue in the office staff. Attention should also be given to the various environmental factors that contribute to the comfort or discomfort of the room occupants: for example, air temperature, air humidity, air flow and speed, radiant temperature, and the general levels of tidiness and cleanliness. Noise should be kept to a minimum: for instance, floors can be carpeted or noisy equipment housed in booths. The room layout should be such as to aid, rather than hamper, the efficient performance of tasks. And there should be provision for adequate maintenance of equipment and the workplace.

SUMMARY

This chapter has emphasised that systems design is a multi-layered, multi-faceted phenomenon. It operates in 'broad-brush' strokes on the overall system, and then with ever finer detail on a host of lower-level systems ('subsystems').

Some design activities are the direct responsibility of the designated in-house staff or appointed external specialists (consultants). Other aspects of detailed design (eg the organisation of logic circuits in a computer central processing unit) are a matter solely for the design engineers of the manufacturer or supplier. Following the carrying out of a feasibility study, specific systems recommendations are made available in a carefully-structured report. When senior management, having considered all the relevant (technical, social, organisational and commercial) factors, decide to proceed with the recommended plan for office automation, the various defined design activities are set in motion. A start is made on the task of acquiring suitable equipment (Chapter 6), staff are retrained or appointed, and any necessary changes made to the office environment.

Systems design, at whatever level, should only proceed after careful analysis of all the relevant factors. It should be recognised at the outset that incompatible requirements will be encountered (demanding optimum compromises), that mistakes will be made, and that not all in-house staff and external users of the office services will be satisfied by the changes that office innovations bring. An effort should be made to design flexible systems that can adapt to new circumstances and improve the services that they exist to provide.

6 Choosing Office Systems

INTRODUCTION

The design of the overall office automation system entails recommending and selecting appropriate equipment and software. The selection of such items can only follow the completion of a satisfactory feasibility study (see Chapter 5). The study includes the recommendations on the type of equipment that will best meet the requirements. And the study should also equip management to approach equipment suppliers (see Vendors, below) and to become usefully involved, where appropriate, in the practical assessment of office system performance, using benchmarks and other techniques.

After the report of the feasibility study has been perused, management should have detailed knowledge about the range of office services that are envisaged, and about the associated staffing and equipment requirements. The needs of the relevant staff – managers, technicians, secretaries, typists, etc – will have been evaluated, and management will have taken into account future workloads in the light of company expansion (or contraction).

The feasibility report will include a requirements specification that can be submitted, with an invitation to tender, to a selected list of equipment suppliers. When the suppliers have submitted their tenders, the intending purchaser will choose according to an assessment of many relevant factors: eg price, maintenance facilities, the availability of training, the reputation of the vendor, etc (see Evaluating the Proposals, below).

The equipment selection procedure, beginning with an initial

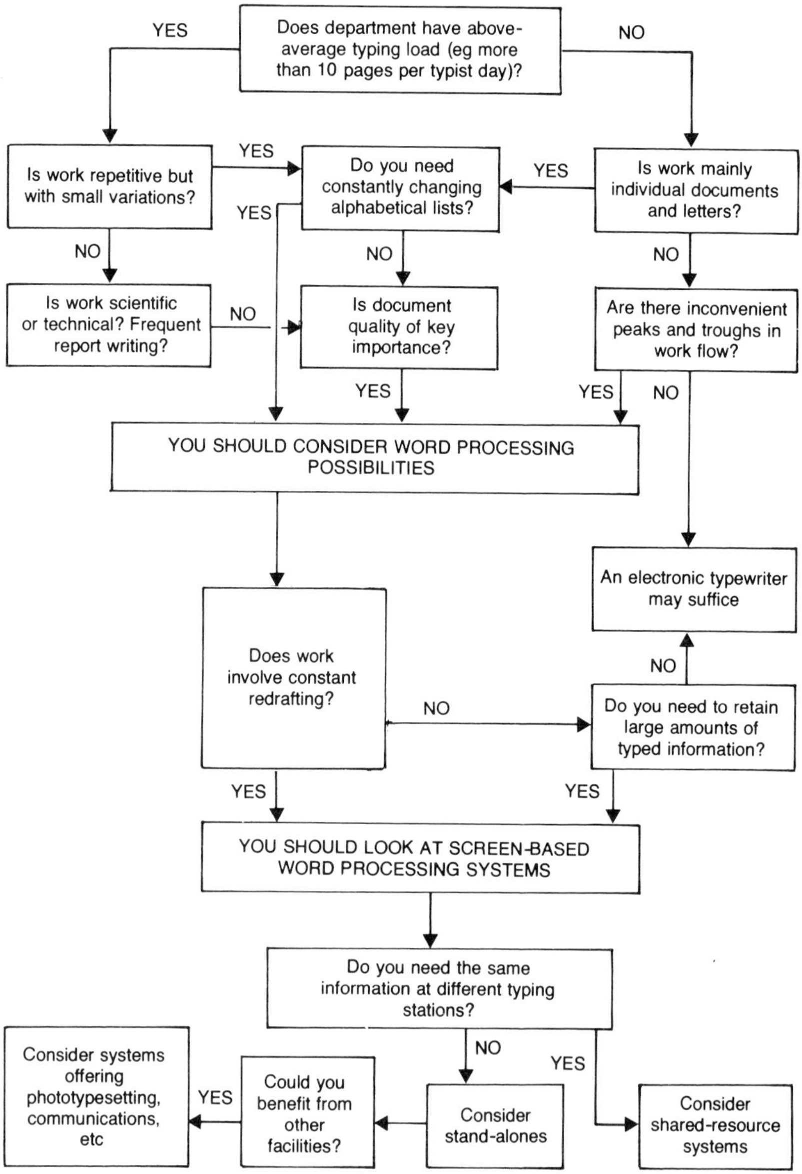

Figure 6.1 Stages in Assessing Need for Word Processing

investigation (in the feasibility study) of office workload, can be shown as a flowchart. Figure 6.1 illustrates this procedure in connection with the acquisition of word processing facilities: the same procedure can be applied, *mutatis mutandis,* to the acquisition of any particular type of office system.

It is important for managers to appreciate exactly what they are getting in acquiring an office system. Does the deal include 'consumables' (disks, stationery, etc), back-up facilities, maintenance, training, etc. The precise terms of the agreement must be clearly specified in the contract (see below).

COST FACTORS

Some of management's most important deliberations will focus on matters of cost, profitability, etc. For example, the prices of available items of office equipment will be considered carefully against performance, availability, capacity for expansion and other features.

In recent years total systems costs have been falling, but not as rapidly as have the costs of pure processing and of main memory. Many peripherals and terminals have remained at roughly steady price levels, but their capabilities have improved enormously, so they are offering ever better value for money. It has always been expensive to develop software, but a growing range of software engineering techniques and 'off-the-peg' software packages have helped to reduce the costs of software development. Telecommunications facilities represent a rising cost as these elements become a larger part of the total systems costs (telecommunications tariffs vary greatly in different countries). And various non-electronic items (packaging, power supplies, cables, etc) represent an increasingly large proportion of the hardware price.

Managers should be aware that many elements in the overall system configuration will have to be purchased separately. Services such as documentation and training may be charged separately, and hardware and software components are often marketed as discrete modules – so that a user can 'tailor make' a system, acquiring only those items that are specifically needed for the required system.

The rapid pace of technological change often makes it difficult to know when precisely to invest in a system. It can be tempting to delay a decision just a few months to exploit the newest technology: there is always a better system 'just around the corner'. But if the feasibility study has recommended a particular type of system, then the sooner the company begins to reap the benefits of well-conceived investment the better. And seductive new systems, untried in the working environment, may not always deliver the imagined benefits.

The economic aspects of the feasibility study will have focused on the costs of the current system, the costs of the 'ideal' solution, and the costs of the available alternative systems. The selection of a system should not just depend upon hardware and software costs: costs associated with system conversion, staff retraining, etc should also be taken into account. It is useful to list, partly by way of summary, some of the important cost considerations that should influence the choice of an office system (this list follows Peltu, 1981):

— the costs of business *in*efficiency, such as excessive stocks, poor cash flow, lost invoices, fraud, consumer dissatisfaction, lack of accurate and timely information on which to take management decisions, slow response to enquiries, etc. Computers can overcome *or* exacerbate such costs;

— management and staff costs taken up in helping to develop the system and in training people to use it;

— systems development costs (designing, programming and testing), which can often outweigh the costs of the actual system. Feasibility study estimates of design costs can be over-optimistic: unforeseen problems may emerge during design, and staff shortcomings may also emerge. Software packages can cut systems development costs, as can such techniques as structured analysis and programming;

— conversion costs, particularly when moving from one computer system to another, could block the way to a desirable solution. Costs are incurred in the parallel operation of the old and new systems, in establishing computer files for the first time, in converting software to run on a new system, in redesigning the stationery, etc;

— personnel costs, which are likely to include the use of external consultants and 'contract' systems, programming and operations experts who work for a defined contract to overcome a period of peak development, to satisfy a particular skills requirement or simply to overcome staff shortages. A shortage of skilled computer professionals also means that recruitment and training costs will remain high;

— equipment costs, which extend beyond the main computer items, include ancillaries and supplies, such as magnetic tape racks for tape libraries, the creation of a suitable environment for the computer-based systems (which are likely to require a clean, air-conditioned atmosphere), stationery, printer ribbons, security devices, etc;

— the cost of failure. If a breakdown occurs it should be possible to resume operations quickly. An unreliable system can be fatal to some organisations. Repair and maintenance facilities are likely to be costly: they are always essential;

— the profit of success. An efficient computer system will not only provide an improved way of performing previously non-computerised activities but will create new services and opportunities and assist the quality of management decision-making.

These various cost factors should be assessed against the various ways of gaining access to computer-based systems (buying, renting or leasing your own in-house systems; accessing bureau facilities; using an in-house system which is owned and run by a service company).

VENDORS

Details of vendors of office systems can be obtained from the trade press or from publications such as *The Computer User's Year Book* in which listings are provided. Some suppliers deal with both hardware and software, others with (predominantly) one or the other. Sometimes a supplier – dubbed 'turnkey' – will supply a comprehensive package of software, hardware and support. And some computer companies, recognising the growing significance of the electronic office, have started offering office systems (eg word processing packages) for computer configurations sold primarily for data processing.

Vendors of office systems sometimes offer a free survey service prior to taking an order. Such pre-sales studies of a client's needs can hardly be regarded as totally disinterested since the vendor obviously hopes to secure an order. It should never be assumed that such a survey will serve in the stead of the customer's own feasibility study. However, a vendor's survey may indicate points to be considered and questions that should be investigated in more detail.

A company may decide to grant an independent consultant the remit of selecting its office system. Here it is important to be certain that the consultant is genuinely independent, ie that he has no connection with any equipment vendor. And such an independent consultant may also be asked to assist in the installing of the new equipment, in planning the new procedures, and in advising on staff training and other matters.

The potential client usually carries out the feasibility investigation and then approaches the vendors. Sometimes, however, a representative of the vendor approaches a company that may or may not be contemplating installing office systems. Here the company may learn about new systems that may be of benefit – or it may find itself purchasing equipment for which it has no real need.

The relation between the vendor and client is obviously crucial to the adequate marketing and successful implementation of office systems. If salesmen effectively 'oversell' the capabilities of systems then short-term commercial gains by manufacturers will quickly evaporate. It is obviously important for all concerned that vendors accurately describe their products and that businessmen and managers take the trouble to assess the real possibilities.

Suppliers may variously offer:

— special terms for new customers;

— 'end-of-line' special offers;

— sales promotion and commission premiums for new products;

— cash settlement discounts;

— 'free' documentation, training, maintenance, etc.

It is up to the manager or management team, possibly with the help of the appointed consultant, to chart a path through the various competing offers.

GOING TO TENDER

When the company management, having studied the feasibility report, decides that computer-based office systems are a worthwhile option the next stage is to prepare a tender document upon which the suppliers can base their proposals. A list of suppliers should then be drawn up and a copy of the tender document, with a covering letter, sent to each. (A sample tender document is given in Appendix 3, a sample letter to suppliers in Appendix 4.)

The potential client company may have its own office-automation specialists. They will know about likely suppliers to whom the tender document should be sent. The non-technical manager or first-time user of electronic-office systems can collect information about potentially suitable suppliers from various sources:

— information services, such as those provided by NCC (Oxford Rd, Manchester; 061 228 6333);

— publications, such as the many newspapers, journals and books aimed both at the computing specialist and general management. Again NCC can provide advice on publications dealing with office automation;

exhibitions and conferences which provide an overview of current industry trends and suppliers. Some exhibitions and conferences focus specifically on information management or the electronic office (some focus on specific office systems: mailbox, word processors, etc);

— word-of-mouth and talking to other users. Existing users may be regarded as the most effective source for discovering the reality behind the glossy sales claims. There are many informative user groups built around a particular supplier's equipment or particular applications (again, contact NCC).

Information should be collected from the types of sources listed above, before any particular suppliers are invited to tender.

The tender document (Appendix 3) should give the prospective suppliers an insight into the size of the company and its expected growth rate, its type of business and the applications to be computerised. The tender document should also guide the prospective suppliers as to how they should submit their proposals and in what detail. The tender document may be entitled 'Specification of Computing Requirements' and it should contain specific well-defined sections.

The *introduction* to the tender document should indicate why a computer system is being sought, and also provide an outline of the company: type of business, location, turnover, size, expected growth rate, etc. The overall format of the report should be described and an outline provided of the type of solution that is required. (If a turnkey solution is required, this should be pointed out.) The introduction should also indicate if more than one supplier is to be involved.

In a *computer applications* subsection of the document, the requirements are specified for each of the applications proposed for computerisation (it is useful to include details of volumes, frequencies and files, etc in appendices to the document). The proposed implementation plan should be outlined, showing the order in which the office systems (if there are more than one) are to be introduced. This should allow initially stand-alone systems to be integrated with other systems as the implementation plan proceeds. Possible enhancements and special requirements should also be detailed. It should be made clear on what systems and facilities the supplier is expected to quote.

A *supplier's proposal* subsection should be used by the suppliers when they are preparing their proposals. Here the suppliers are advised as to the section headings they should adhere to when preparing their proposals. The aim is to standardise the layout of the proposal documents, so that later comparisons between proposals can be easily made. Section 3 of the sample tender document (Appendix 3) itemises the details that the supplier should include in the proposal.

Appendices are required to specify the volume of transactions for each application. If possible, an indication should be given of the item details maintained on each record within each file, showing the number of characters in each item. A guide to record content and number of characters is useful for suppliers trying to estimate file storage requirements. An effort should be made to indicate the detailed requirements of each proposed office system. Only with this approach can ambiguities and later confusion and disagreements be avoided.

Having received the tender document (and covering letter) the suppliers should be expected to submit their proposals by a specified date. When the prospective client receives the full batch of proposals, they are then systematically evaluated before a decision can be taken and a contract, with one or more suppliers, agreed.

EVALUATING THE PROPOSALS

Before any direct effort is made to evaluate the suppliers' system proposals it is useful to consider various points for each supplier. For instance, an attempt should be made to ascertain the financial stability of the company and how long the company has been operating. How much of the supplier's business is devoted to the type of system recommended? How many of the proposed systems are installed in this country and worldwide? How many staff are employed by the supplier, devoted to support services, and specialising in relevant application areas? Do service centres and relevant user groups exist? Where are the users with similar systems? When this sort of information has been gathered, it is possible to start the detailed evaluation of the various supplier proposals. The aim is to ensure that the proposed systems can meet, in a cost-effective way, the company's requirements.

There are many ways of approaching the evaluation task. Some are based on a points-scoring system, others on a financial-scoring system. One of the most effective methods is the 'weighted ranking' analysis, in which each item is awarded points against a list of predefined criteria which have been weighted according to their importance. Before the detailed weighted-ranking analysis begins, the prospective client may be advised to calculate all costs over a five-year time space: here attention should be given to the costs of hardware, software, training, documentation, delivery, installation, maintenance, etc. If it is found that particular costs or the likely total investment are unacceptable then there is little point in embarking upon a detailed analysis of systems features.

Where the total costs are within the company's allocated budget, the analysis can proceed in a number of well-defined steps:

— *identify criteria for the evaluation*
  There are various ways of grouping the criteria. We may opt for broad headings such as Hardware, Software, Support and General (or Functional, General and Human Aspects). The requirements of the supplier should determine the headings that are selected. Specific criteria should then be listed under each heading;

— *identify essential criteria*
  If a piece of equipment does not satisfy any one of the *essential* criteria then it is unsuitable and should not be considered further. The identification of essential criteria can be used to draw up a short-list of candidates for more detailed scrutiny;

— *assign weightings to the remaining criteria*
  The weights can be assigned as a percentage to the various criteria. The percentages allocated to criteria that can be grouped together should add up to 100. A weighting matrix is constructed and the various criteria are compared with each other. For example, if criterion A is more important then B then score 3; A equally important to B, score 2; less important, score 1. (Different companies will rate the criteria differently according to requirements.) Each level in the hierarchy of criteria is assigned weights in the same way. (The procedure is discussed in more detail in Beaman, 1983, Chapter 5; and in Damodaran et al, 1980);

— *identify factors for awarding scores to criteria*
It is necessary to list the various factors which are to be judged for each criterion. One suggestion is that an Evaluation Sheet be made out for each feature to be evaluated, showing the feature, the weight to apply, and the names of the suppliers. Scoring using a *sliding scale* is useful where some kind of subjective judgement has to be made; but *fixed scales,* deriving from unambiguously quantified features (eg price, size, etc) are usually preferred;

— *evaluate each proposed system*
Each submitted proposal can now be analysed to ascertain how well it satisfies each criterion. For each criterion, the potential client should decide which system factor is most relevant. A score is allocated and the next criterion is then considered. When all the criteria have been dealt with, each score is multiplied by the weighting for that criterion. Finally the results are totalled for each type of criterion;

— *compare results and make selection*
Once each proposal has been evaluated, the system with the highest score wins. If several models have high scores which are close, the whole procedure can be repeated using new, more refined criteria; or a panel can be assembled to choose the system.

AGREEING A CONTRACT

A carefully drawn-up contract between the supplier and the user of electronic office systems is essential to the future operational and commercial success of the systems. The best system may have been selected by the client company, but it may be a failure if unexpected contingencies occur or unexpected costs accrue. The contract should aim to define the terms of the client/supplier agreement so that responsibilities (and cost liabilities) are unambiguously defined. Senior management should oversee the formulation of a contract for the acquisition of office systems.

A standard contract made available by the supplier may be expected to be biased towards the supplier's interests. The client, with different interests, should make sure that the rights and obligations of *both* parties are made explicit in the contract. Key aspects of the systems specification (developed during the feasibility study) should be included in the contract. The complexity of computer-based systems makes such an inclusion more, rather than less, necessary. Management should obtain advice, if necessary, on the form of the contract. The Institute of Purchasing and Supply, for example, has developed a model contract which gives attention to the following points (and to others):

— all key specification criteria developed during the feasibility study should be quantified and summarised;

— any doubtful terms within the contract should be clearly defined (in the contract);

— the contractor must accept responsibility for examining the environment in which the system will operate. It should be stated in the contract that the contractor knows the conditions under which the system will be expected to operate;

— the responsibility for extra costs that occur due to the supply of incorrect information must be explicitly attributed to the party at fault;

— equipment or software should not be delivered without the customer's permission;

— procedures should be established to govern any contractual variations required by either party;

— the responsibilities, if the system fails to meet the performance criteria, should be specified. The client may want additional equipment, a price reduction, or damages if there is manifest breach of contract;

— the customer should be able to obtain recompense from the contractor if the performance of the system deteriorates during live operation after the acceptance tests, if the deterioration can be shown to be the responsibility of the supplier;

— the customer should be protected if the supplier goes bankrupt;

— information provided by either party during the project

should be protected by a confidentiality clause;

— the position should be clarified when equipment and supplies from other vendors are used. The client should be able to use other vendors, but the main supplier should be protected from extra costs that may arise;

— the supply of hardware spares during the lifetime of the system should be protected as far as possible;

— software products and services should be protected. For example, software and documentation necessary for the operation of the system must be provided with the equipment. The customer should reserve the right to withhold payment for any aspects of the system for which adequate documentation has not been supplied.

A sound contract can minimise disputes and confusion between customer and supplier. For this reason, the clearest possible definition of rights and obligations is essential.

In addition to contractual protection, the customer should also seek protection through insurance. Here the amount to be assessed for insurance cover is likely to include material damage and any loss to the company through system disruption. The insurance company will want to know if the company has back-up provisions, secure copies of software and documentation, and provisions for reconstituting the service quickly. (Addresses of likely insurance brokers and insurance companies can be obtained from Lloyds of London.)

SUMMARY

This chapter has focused on how a company should approach the task of system acquisition. It has been emphasised that various cost and profitability factors will be key influences in decisions to acquire systems and services from particular suppliers.

The feasibility study yields a requirements specification which can be used to frame a tender document to be sent to selected suppliers. The subsequent supplier proposals can be evaluated according to weighted-ranking methods or other forms of analysis. The specification produced during the feasibility study can then be used as an element in the contract with a selected supplier (or suppliers). Attention should also be given to the question of customer protection through insurance.

Part 3

The Human Factor

7 Managing People in the Office

INTRODUCTION

There is a growing recognition that adequate attention should be given to the roles of people in systems. Computer-based equipment and software do not, alone, define an overall office system. The people involved should also be regarded as important system elements, not to degrade people to mere 'cogs in the machine' but to encourage the levels of morale and motivation without which the system cannot realise its full potential. The new appreciation of the significance of people in functional systems is signalled by various phrases in the literature: 'human factors', 'human sciences' (as in Human Sciences and Advanced Technology, HUSAT, of Loughborough University), 'the people aspect', etc.

Some of the broad 'people considerations' relate to the question of employment. What are the likely consequences of electronic-office systems for staffing levels? In the short term? In the long term? More specifically, systems design should aim to maximise motivation, to ensure that people work in an environment that is both psychologically rewarding and physically healthy (see Chapter 8). Considerations of human factors should influence how systems are configured and operated. Salesmen sometimes boast of how easy it is to operate their systems, how simple (for example) are the function keys on the terminal. But such claims should be assessed with care. Staff are likely to become bored and apathetic if systems are too simple to operate, too automatic, too impersonal. Operators have been known to introduce deliberate errors into their work to make their activities more interesting! System design should aim to discourage this sort of possibility.

There are several 'human-requirements' areas, in addition to matters of systems design. We can see that human factors are clearly relevant in staffing levels (the employment question), in industrial relations (the roles of trades unions, technology agreements, etc), motivation and morale, training and organisational structure. When any of these (and related) areas are ignored when an organisation seeks to introduce office automation, problems are likely to arise. At best there may be a loss of productivity, the system working below predicted levels; at worst the system may be totally shut down through sabotage or industrial action. Managers should take human factors into account as a means to achieving operational efficiency and business success.

PERSONNEL MANAGEMENT

We would expect new office technology to affect every aspect of personnel management. The management policies used to handle personnel matters need to be sufficiently flexible to adapt to the changes in office life brought about by electronic-systems innovations. It will be necessary to consider, and possibly revise, the policies in such areas as:

— job evaluation and grading;

— career development;

— manpower planning (including redundancies, redeployment and recruitment);

— wages, salaries and other forms of remuneration;

— hours of working and other conditions of employment;

— health and safety;

— labour relations and industrial democracy.

The impact of new office technology will be felt in all these areas. It is important for managers to develop appropriate strategies that anticipate possible difficulties, rather than managers having to adopt ad-hoc, ill-considered expedients when problems arise.

THE EMPLOYMENT QUESTION

It is assumed that managements take decisions which they believe will benefit their companies – in terms of increased market share, prospects for advancement into new product fields, etc. Very often, company interest is synonymous with employee interest, but not always. It is inevitable that management decisions will sometimes cause problems for employees (and potential employees). This is particularly so in the area of job security: simply put, improved productivity can be achieved by altering the ratio between staffing and production – such that, for example, the same number of staff do more work, or fewer staff do the same work. In a buoyant national economy, it is quite possible for companies to take on extra staff and still to improve productivity through expansion in production. But it is hard to avoid the conclusion that in many of the industrialised countries, Britain included, improvements in productivity generally mean either redundancies or a block on further recruitment (so affecting the prospects of job seekers).

Many studies have suggested that office automation will have a significant impact on employment in the years to come (as other forms of automation will have in other industrial and business sectors). For example, in one study (Bird, 1980) it was found that two word processor users achieved productivity gains of one hundred per cent or more. A steady growth in the number of word processors was predicted, with an estimated job loss by 1990 of 170,000 typing and secretarial jobs in the UK, seventeen per cent of the total. And we may speculate that other types of office automation will have comparable effects on employment in the office environment. At the same time it is clear that the new electronic office systems will create jobs, both in the provision of the systems for customers and in the new facilities and services generated in the office itself. In any event, company managements will want to interest themselves in staffing matters. They will give attention to such questions as new skills requirements (involving training and recruitment), age considerations (younger people may be more responsive to changes to the office involving computer-based systems, and plans should be developed to help older staff who may have difficulty in adapting to new ways), and regional differences in staff resources (organisations may choose to relocate some offices nearer to pools of skilled staff).

Computer-based systems are having a growing impact on employment patterns. Some jobs are becoming less frequent, others are emerging for the first time. It is hard to predict what will be the overall consequences for staffing in tomorrow's office, but it is obvious that both managements and staff will need flexible attitudes to cope with the accelerating pace of change.

TECHNOLOGY AGREEMENTS

Technological change has always influenced the relationship between 'managers' and 'workers'. Perhaps most significantly the Industrial Revolution led to the development of urbanised working classes that eventually formed the industrial trades unions (and later the British Labour Party). The emergence of the trades unions in the 19th Century led to efforts to protect manufacturing employees from the adverse effects of new technology. More recently, specific 'technology agreements' have been negotiated, in various countries, between trades unions and managements.

In the 1950s efforts were made to develop a sociotechnical approach to work design, in which appropriate attention is given to the various human, organisational and technical needs. In Scandinavia, for instance, new working methods were created in which the *group* – containing a variety of skills and performing a range of tasks – was regarded as crucially important. This development, and parallel trends in other parts of Europe and the United States, encouraged trades unions to demand a say in corporate decision-making. There was much discussion about industrial democracy and other modes of worker involvement.

Specific *technology agreements* (sometimes called 'new technology agreements') were developed in Scandinavia in the mid-1970s to cover computer-based information systems. It was seen that such agreements fitted easily into the labour-relations environment of particular countries – with established traditions of industrial democracy, strong national bodies representing unions and management, and a concern for working conditions. The new technology agreements provided a model for white-collar unions throughout developed countries in the western world: supported by comprehensive union research the new agreements suggested how workers could be protected from the adverse effects of innovative computer-based systems.

In 1979 the British Trades Union Congress (TUC) published a booklet dealing with technology and employment. It included a ten-point plan intended to serve as a basis of negotiations for technology agreements. The main TUC recommendations are that:

— no new technology should be introduced unilaterally by management; a status quo clause should be included to require agreement to be reached between management and unions *before* any new technology is introduced;

— unions should have guaranteed access to information about management objectives for technological innovation at an early planning stage, before any decisions have been taken;

— there should be no redundancies as a result of technological change, or there should be good redundancy payments; plans should be developed for staff redeployment and relocation that may result from the technology;

— adequate provision should be made for necessary staff training and retraining, and redundancy notice periods should be used for retraining;

— unions should have a say in systems design and reprogramming; procedures should be agreed to cover personal data held in computerised form; and computer gathered information should not be used in work-performance measurement;

— stringent standards should be established on health and safety;

— working hours should be reduced and systematic overtime eliminated to spread available work amongst more people (this assumes that technology will improve productivity and probably lead to a general increase in unemployment);

— unions should build up expertise in relevant technologies; 'technology representatives' or 'data stewards' should be appointed to represent staff interests in technical negotiations and there should be inter-union collaboration in negotiating technology agreements;

— technology should not disrupt existing pay structures;

— a joint management/union team should monitor technological developments.

We may expect managements to dispute a number of these points. However the TUC guidelines indicate the implications of information technology and highlight key areas of industrial relations.

Four main ways of controlling negotiations about technological change can be identified (Figure 7.1 gives examples of these types of agreements):

— *laws and regulations,* particularly regarding industrial democracy and health-and-safety provisions;

— *national framework agreements,* negotiated between national management and union bodies;

— *sectoral agreements,* covering a particular industry or type of activity;

— *local agreements,* relating to particular organisations and workplaces, with some corporate framework agreements covering an organisation used as a guideline for negotiations within subsidiaries, divisions and workplaces.

It has been suggested that the crucial agreements are the local ones. Some organisations have found different experiences in different divisions covered by the same framework agreement. As with all union/management negotiations, it is inevitable that new-technology considerations will be interpreted according to the perceived interests of the parties. Managers have argued, for instance, that the TUC recommendations can delay system development, allow new technology to be used as a lever to achieve other demands, delay moves to improved productivity, and damage the manager's legitimate right to manage. Whether such fears are justified will depend upon the particular management style, the union expectations, and the prevailing industrial-relations situation in the particular working environment. It should be emphasised, however, that systems design in which there is participation by representatives of system users is more likely to result in a

Country	Laws and Regulations	Collective Agreements		
		National	Sectoral	Local
West Germany	Works Constitution Act 1972; Works Safety Act 1973; plus VDU regulations 1981	None	Job protection in metalworking, printing, textiles, leather and other industries	Over 100
UK	Health and Safety at Work Act 1975 plus VDU guidelines 1983	None	Parts of public sector	Over 200
Norway	Working Environment Act 1977 plus VDU regulations 1982	Employer/Union agreement, 1975	Banking	Most workplaces covered
Sweden	Co-determination Act 1977; Working Environment Act 1978 plus VDU regulations 1981	Employer/Union Work Environment agreement 1976	Technology agreement in printing; codetermination in many public and private sectors	Use of legislative rights
Various	Health and Safety laws, eg in Italy and France; codetermination laws, eg in Austria; Statute of Workers Rights 1970 in Italy	1981 agreement for private sector in Denmark	Printing sector in Belgium, Holland, Austria and Greece; metalworking in Italy	In USA, 1979 Ford agreement; General Motors quality of working life programme; in Japan, job security in big companies in return for union loyalty

Figure 7.1 Examples of Procedures for Joint Regulation of Technological Change

(Source: chapter by John Evans of European Trade Union Institute in *New Office Technology: Human and Organisational Aspects* edited by Harry J Otway and Malcolm Peltu, Frances Pinter, 1983)

system that is relevant to real needs. The TUC recommendations should be regarded as establishing a realistic negotiating framework which can contribute towards the realisation of effective computer-based systems to which managers and other employees are committed.

MOTIVATION AND MORALE

Adequate attention to 'people factors' is essential to satisfactory system design, implementation and operation. Where a system has bad effects on users and other involved people, there will be various undesirable consequences; for example:

- difficulties in implementing and operating the system;
- user indifference and recalcitrance, leading to bad industrial relations, staff turnover, absenteeism, sickness, etc;
- use of the system in inappropriate ways and failure to use some facilities;
- deliberate sabotage to make the work more interesting or to relieve psychological tensions.

Where employees are forced to adopt systems and working practices that go psychologically 'against the grain', there will be adverse consequencies for office efficiency, productivity, business success, etc. Good morale is essential to the efficient operation of systems. Reitz (1980), for instance, identified various internal 'motivators' that are relevant to human performance in the working environment:

- a sense of formal responsibility;
- a sense of involvement in the work produced;
- a sense of achievement;
- a sense of recognition.

It is suggested that where managers pay attention to such factors, in the office environment or elsewhere, staff will be motivated and productivity will be maximised (Reitz: 'Cover these bases well. The payoff in terms of morale and productivity is well worth any extra effort put forth by a word processing team and management'.

And what is true of word processors is true also of other electronic office systems).

Many factors are relevant to motivation and morale in the working environment. Job satisfaction, for instance, can only be realised if a number of elements are taken into account. Discussion has focused on financial remuneration, perquisites, working hours, incentive schemes, health-and-safety factors, social contact, work rhythm, involvement in the work, achieving objectives, pride in providing a service, the stimulus of a challenge, the opportunity for creativity, the exploitation of skills, the assigning of status, etc.

New office technology can be used to improve all these aspects – or, introduced by careless and ineffective managers, to damage them. It is hardly surprising if employees resist office automation where it is perceived as resulting in de-skilling, 'Big-Brother' monitoring of performance, loss of personal autonomy, lowering of status, and reduced social contact. (Peltu, 1984, discusses the question of job satisfaction in more detail.)

Many studies have highlighted the factors associated with high levels of job satisfaction. We see frequent reference to the importance of allowing people to develop new skills, to feel that the job is important, to exercise responsibility, to find variety and inherent interest in the assigned tasks, and to liaise with other staff in mutually supportive roles.

There are many legitimate worries about the possibly adverse consequences of introducing office automation. Managers should pay attention to such anxieties as relevant to productivity and business success. The key factors that are relevant to job satisfaction appear self-evident, when actually spelt out. Too often, however, they are ignored by managers charged with the responsibility for introducing office automation.

THE NEED FOR TRAINING

In many fields concerned with the introduction of computer-based systems, there is a shortage of skilled staff. This is certainly true in the area of office technology. Here the shortage of adequately trained personnel could represent a serious limiting factor on the growth of office automation. There is a shortage both of the technical staff needed to design and develop the new systems and of the operating staff needed to use the systems once they are installed. It is also significant that the technology is changing very rapidly: staff trained even a short time ago may now need retraining.

Systems design is relevant to the ease with which managers and staff users can be trained to operate the new equipment. For example, dialogues with the users should not be conducted by the software in jargon-ridden phrases: systems design should, at all times, be as user-friendly as possible. And use can also be made of Computer-Assisted Learning (CAL) software to teach particular skills. A CAL system presents information to a student, and then asks questions about it. If the student responds correctly, the software gives the next lesson; if the student answers incorrectly, revision material is presented until the student finds the right answer. CAL systems are particularly effective training aids when they are used in conjunction with traditional educational methods and experienced supervision.

The new office technology requires the development of various skills in staff with different responsibilities in the organisation. It is worth indicating the relevant skills in 1) secretaries, typists and clerks, and 2) managers.

Secretarial, Typing and Clerical Skills

The new types of office automation may be expected to reduce the amount of routine work carried out by staff at this level. In consequence, training and retraining for the exercise of new skills will become increasingly necessary. For example, there will be an opportunity to participate in the design of systems to be used by secretarial and clerical staff, to supervise small computer-based installations, to assist in the developments of systems and software, to evaluate new systems, to train new staff, to exploit information networks, etc. To assume these types of responsibilities it will be necessary for staff to understand, albeit in broad terms, the principles of the new office technology; to develop an awareness of the overall information system; to understand how to exploit their own systems; and to know what to do when something goes wrong.

In organising schemes for training and retraining office staff, managers should be discriminating in deciding what training to offer to which people. For example, not every typist is willing (or able) to switch to word processing. Perhaps typists should only be selected for word processor training when they have shown appropriate aptitude and involvement. The typists should already have had some involvement in selecting the new systems – they should have seen the equipment prior to the finalising of the contract, they should have tried out the keyboards and screens, etc. By the time a system is installed, management should already have some insight into which staff are most likely to benefit from the various types of available training.

Management should decide which staff are to be trained, what form the training should take, how long it should last, and how existing office services are to be maintained while training is in progress. The answers to all these types of questions will be determined by the size of the organisation, the range of office services that are to be provided, what the plans are for expansion, how staff feel about the new developments, etc. Managers must evolve effective plans for their own situations.

Management Skills

There is a growing requirement for managers to be able to handle the process of technological change in particular departments and companies as a whole. At the same time they need to be able to cope with resulting changes to their own jobs. There are various key management skills for which training plans must be provided.

It is necessary, for example, for managers to understand information technology principles and capabilities. For this purpose, managers must learn about the information technology industry and understand the broad principles of systematic systems design and analysis. They should know how to manage the change process, and develop their competence in organisation, work and job design. It is necessary also for managers to be able to assess aspects of ergonomic design, and to be able to cope with databases and other information resources. Managers will be expected to use electronic mail systems effectively and to exploit computer-based decision-support systems. And they will be able to use teleconferencing provisions, facsimile, word processing, and other electronic-office innovations.

It is obvious that the various management skills are essential to the success of any office innovation. The skills of other staff can only be developed and exploited within a framework created by management. It is upon the experience and training of managers that the ultimate success of any office-automation system depends.

SUMMARY

We can emphasise again that the effective management of people is essential to the success of electronic office systems. If people are not well motivated in the working environment, they will lack commitment, be discouraged from taking creative initiatives and largely indifferent to the success or failure of the computer-based office systems. The sound management of people makes good business sense.

Managers should remember that progress towards the electronic office:

— involves much change in employment patterns;

— affects the job-satisfaction expectations of white-collar staff;

— involves paying attention to job design as a key element in system design;

— involves inevitable tensions and conflict, often of a 'political' nature, both within departments and from one department to another;

— highlights the shortage of relevant skills, so making essential a systematic scheme for training and retraining.

It is inevitable that the introduction of office automation will cause dislocation in working patterns and threats of various sorts to the positions and status of existing employees. As far as possible, the situation should be turned to advantage: change should be seen as a means to improving existing services and to generating new ones; seeming threat should be translated into challenge and opportunity. Change should only follow involvement of staff in systems design and appraisal, due attention to the views of staff representatives, and the negotiation of suitable technology agreements (and, where appropriate, agreements of other types).

Many of the principles that should be observed in the management of people in the automated-office environment are age-old: employees have always wanted (needed) to be treated seriously, with adequate attention paid to their human requirements for satisfaction, respect, stimulation, social intercourse, etc. Some of the principles are newly emerging, generated by the unprecedented features of computer-based office systems. These concern, amongst other important aspects, the health of people in the electronic office.

8 People and Health

INTRODUCTION

One of the most important factors concerning people working in the office environment is their health. Working conditions have always been relevant to the health of employees, and this is no less true where computer-based office systems are being introduced.

In offices, as in all working situations, a number of health hazards can be identified: for example, various correcting fluids can give off toxic fumes; typists and terminal operators can suffer from backache and neckache; strip lighting can cause problems for epileptics and migraine sufferers, etc. And display screens, common in the electronic office, can cause other difficulties. A visual display unit (VDU), like some domestic television receivers, can cause eyestrain, headaches, fatigue and nausea. It has been suggested that, in some circumstances, display screens may cause migraine and epileptic fits. And another idea is that since the units give off low-level radiation they may be responsible for cataracts. There is no evidence to support some of the more alarmist claims, but it can be assumed that suspicion about such possibilities will be bad for morale. It is obvious that the health of staff required to use screen-based electronic systems should be monitored with care. There is debate as to how significant the various effects can be.

In the early-1980s there were fresh fears about the effects of working with display screens. These followed the appearance of a face rash among operators of VDUs in Norway. The possibility of occupational dermatitis was suggested, and some discussion focused on 'photon sensitivity' as a causal factor. It was also pointed out that about four per cent of all epileptics are known to be photosensitive. Another incident at about the same time concerned the birth of four handicapped children in Canada. The mothers had worked for lengthy periods on VDUs at a Toronto newspaper during their pregnancies. Subsequent tests confirmed that the terminals were not emitting extensive X-rays or microwave radiation that might damage a foetus.

A 1981 report commissioned from the Christian Michelsen Institute in Bergen, Norway, described how electrostatic potentials can build up between some screens and earth, and how this can cause some operators to carry their own high charge. This charge can irritate the operator's skin, possibly causing rashes, and can cause eye irritations. We would expect both unions and management to be concerned about such possibilities.

The research evidence from around the world suggests that VDUs *can* cause various health problems – eyestrain, backache, wrist pains, etc – when the systems are poorly designed and inadequate attention is given to the overall working environment (lighting, atmospheric controls, workspace, furniture, etc). This also illustrates a general point: it is essential to take human factors into account in designing office systems of all types, not least because system design is relevant to the health of the human users. (See Ergonomics, below.)

HEALTH AND SAFETY

Apart from such aspects as visual fatigue and muscular fatigue (discussed below), it is possible to indicate a wide range of possible hazards in the working environment. Damodaran et al (1980), for instance, consider a number of health-and-safety hazards that are relevant to users of electronic office systems:

— Being struck by
  — sharp edges;
  — pointed corners;
  — falling equipment which was unstable;
  — moving equipment or machinery;

	— moving parts of equipment or machinery (eg in some types of photocopiers).
— Being injured while handling objects	— which are too heavy or cumbersome;
	— which are inconveniently shaped or sited.
— Environmental effects	— toxic or noxious fumes given off by equipment during normal operation;
	— toxic or noxious fumes given off by equipment during breakdown;
	— fire caused by equipment faults or failure.
— Electrical shock	— from 'faulty' equipment;
	— caused by inappropriate action by personnel.

It is emphasised that most of these hazards can be eliminated by appropriate selection of equipment, and proper checking on delivery for faults; and by clearly stating appropriate and inappropriate personnel actions and activities (reinforcing these during training). Aide-memoires (eg 'under no circumstances remove this panel until all power to the device has been switched off') can be used.

Many of the principles that should govern the selection and installation of electronic office systems, in the interests of health and safety, apply equally to any equipment that is to be used by human beings. Other principles have been formulated specifically for the design and use of computer-based facilities. It is useful to consider visual and postural factors in more detail.

VISUAL FACTORS

The effects on users of working with display screens have been studied for many years. For example, Dr G Busch observed in 1976 the 'eyestrain and visual fatigue created by job tasks performed in a working environment which is ill-suited to healthy work habits', noting that 'this is particularly true for those work environments utilising visual display units as an important part of the job load'. He identified what he saw as 'visual fatigue' caused by the use of display screens, and listed various symptoms:

— discomfort of the eye, feeling that the eyes are tense, heavy, dry and uncomfortable, burning and tender, producing 'grittiness' (ocular symptoms);

— bad focus, double vision, blurred edges on figures, inability to look in any one direction continuously, colour fringe on objects viewed (visual symptoms);

— headache in forehead with stabbing pains, muscular aches sometimes spreading to the arms and neck areas (systemic symptoms);

— high absentee rates, increase in days lost to illness (behavioural symptoms).

A variety of other factors can affect the frequency and intensity of the symptoms. Excessively boring or frustrating work may encourage the operators to resort to alcohol or other drugs. People on Valium or Librium may be more likely to react adversely to long periods of contact with a display screen. Alcohol and other drugs are known to make smooth eye movement more difficult and to increase fatigue and eyestrain. A number of studies have found that a substantial minority of VDU workers experience frequent eye trouble of some sort, with as many as 40 to 50 per cent suffering less frequently from similar symptoms. A woman working at France's biggest national statistical office in Nantes has been quoted: 'On an average day we type between 700 and 1000 pages. When we get tired and letters start dancing in front of our eyes, the only way to stop conversation with the screen is to switch it off. But when we do that we lose our productivity bonus'.

The term 'visual fatigue' can be used to describe a variety of symptoms, and the physiological causes of the symptoms are numerous. For example, there are small muscles used to control the focal length of the lens of the eye: these can become fatigued, causing the sufferer pain in the eye, behind the eye, or a more diffuse headache. Sufferers may also encounter an itching in the eye, spots before the eyes, and difficulties in changing the focus of the lens for near and distant objects. There is debate as to the extent to which such symptoms can be caused by prolonged exposure to display screens, but various task- and equipment-related causes of visual fatigue have been identified:

— constant refocusing of the eye to accommodate different viewing distances;

— frequent eye movements near the extreme ranges of available movement;

— continuous focusing at a near point for prolonged periods (more than thirty minutes);

— undesirable aspects of the visual environment, eg excessive glare from over-bright light sources, insufficient levels of illuminance at the workstation;

— eye defects which are wholly or partially uncorrected.

Many of these causes of visual fatigue can be reduced or eliminated by appropriate job and workstation design. In some instances (eg uncorrected eye defects), opthalmic assistance is required. At all times, managements should be aware of the possible causes of visual fatigue, and should arrange for frequent and regular checks on staff.

It has been pointed out that normal reading glasses are often unsuitable for VDU work because they are intended for distances of only 35cm (VDU usage involves viewing distances of between 45 and 70cm). Tests on the eyesight of VDU operators should take this circumstance into account. Special lenses (varafocals, trifocals, etc) may be appropriate in particular cases.

VDU operators may have to scrutinise screens for long periods, and to cope with rapid switches of focus between the screen, the keyboard and documents used as source information. Any attempts to estimate the causes of visual fatigue should take these various conflicting demands into account. And special eye characteristics, such as colour blindness, should also be considered.

POSTURAL FACTORS

There are various postural factors that can cause undue fatigue in office workstation operators (visual fatigue may be an early symptom). The design of chairs and desks, and the positioning of screens, keyboards, source documents, etc should encourage comfortable body movements, discouraging inconvenient stretching and uncomfortable muscular stresses.

Muscular work usually involves alternate contraction and relaxation of muscle tissue, which increases the rate of blood flow through the muscle. Blood vessels in the muscles carry valves: a contracted muscle becomes hard and the blood vessels are drained, and when the muscle is relaxed the blood vessels are filled again. The resulting blood flow, during dynamic work, gives the muscles ready access to energy-releasing material and allows waste products which impair muscle function to be removed. In static muscular work the muscular pump stops functioning and the muscle quickly tires. These circumstances are relevant to the activity of an operator at an office workstation. In broad summary, office systems may cause the adoption of two kinds of undesirable postures:

— *awkward postures,* adopted to compensate for bad workstation design, visual factors and unadjustable furniture;

— *fixed postures,* caused (for example) by fixed keyboards and communication and control facilities that reduce the need for movement. It is recognised that decreased mobility (of the operator) can have adverse physical and psychological consequences.

The general requirements are clear. Systems should be designed to adapt to human physical and psychological needs. We know the factors that cause undue fatigue and that may also represent more serious hazards to workers in the electronic office. Our approach to overall systems design should reflect our awareness of the various potential threats to working men and women. In this way, systems design will take full account of the full spectrum of ergonomic considerations.

ERGONOMICS

Ergonomic considerations are important for the convenience,

comfort and safety of office workers. They are also relevant to the reliability of systems: for example, bad human/machine interfaces can encourage operational errors. Partly by way of summary we can list the various types of factors to which ergonomic consideration should apply:

— visual;

— postural;

— personal;

— environmental;

— task-requirement;

— nature of system.

These factors may be seen to overlap with another spectrum of design considerations, namely that establishing the parameters for effective workstation design. These parameters can, for instance, involve *defining the likely user population,* employing anthropometric data for this purpose (thorough design of workstations can require detailed knowledge of, for example, the range of limb movement for men and women). Other parameters may focus on whether the workstations are to be designed for *sitting or standing operators*. And attention also needs to be given to chair and work surface *heights, viewing distances* from screens, the need for *eye and head movements,* and the provision of *document holders*. (These elements are discussed in detail in Damodaran et al, 1980.)

Personal factors can include such conditions (already mentioned) as uncorrected visual defects, age, health, obesity, and addiction to nicotine, alcohol or other drugs. Environmental factors are concerned with such elements as lighting, heat, humidity, noise and the provision of defined territories for particular tasks and workers.

To give a flavour of the ergonomic quantification of certain important factors, the following recommendations are included (more are listed in Peltu, 1984):

— *chair:* adjustable height (34 to 52 cm, typically about 40 cm); adjustable back support; good stability with base on self-locking castors; rounded front edge to seat and padded seat surface of textured material;

— *viewing distances and head/screen positioning:* eye-to-screen and eye-to-document, 35 to 70 cm, optimally 45 to 50 cm; eye-to-keyboard, 40 to 60 cm (preferably about the same as eye-to-screen/document); eyes cast down at an angle of 15° to 20° to the horizontal; screen inclination from vertical 20° maximum; all items within 25° to 30° range relative to normal viewing;

— *footrests:* adjustable between 4 and 11 cm high and 5° and 15° in inclination;

— *lighting:* room illumination should be between 300 and 700 lux (lx), optimally about 500 lx, with illumination over 700 lx advisable only where there are no VDUs, microfilm or other light-emitting equipment.

Attention can be given, in a similar fashion, to questions of workstation siting, noise, curtaining, wall colours, cabling, keyboard dimensions, display sizes, screen character formation, etc. The important point to remember is that every detail in the office environment can be analysed and quantified: design recommendations can then be made in full awareness of known physiological and psychological characteristics. This thorough ergonomic approach helps to optimise the design of overall systems in which people are key elements. In such a way, systems are made efficient and human beings are protected from a wide range of environmental hazards.

SUMMARY

Increasing attention is being devoted to 'human-factor' (or ergonomic) considerations in designing, implementing and installing electronic office systems. This attention is highly relevant to the health and safety of workers in the office environment.

Today we know much about the visual, postural, personal and environmental needs of human beings in the workplace. This insight, coupled with knowledge about the technical design of systems, has allowed ergonomics – a truly cross-discipline study – to influence systems design to the advantage of employees and organisations alike.

The well-being, physical and psychological, of people in the electronic office is not simply a humane and civilised requirement, a demand of natural justice. It is also a condition of commercial success. The committed employee, safe from environmental hazard, is a desirable element in any business enterprise. A systematic and thorough application of ergonomic principles allows system designers and managers to approach this worthwhile state-of-affairs. To invest in the safety and good health of office workers is a sound business policy.

Part 4
The Future

9 Tomorrow's Office

INTRODUCTION

The shape of tomorrow's electronic office is predictable in part, though we can also expect surprises. We can say, for example, that the amount of paper in use will diminish and that workers will be less restricted to a particular geographical office site. Many companies have now abolished virtually all usage of paper from certain office environments and the overall trend – discernible also in universities, hospitals, defence establishments, etc – is unmistakable. And as paper gives way to the electronic storage and dissemination of information, anyone with access to a telephone line will be able to use (and contribute to) the office services offered by companies and other organisations.

The surprises will come from the new functions that future computer-based systems will be able to perform. For example, little has so far been said about how the burgeoning range of *expert systems* will affect existing office services and yield new ones. In this area, as elsewhere, new technological developments will constantly shape and reshape the relationships between 'workers' and 'management', between employees and their company, between office workers and customers, between citizens and what opportunities and facilities they are offered by their society.

This chapter briefly indicates some of the influential factors that are affecting the evolution of the electronic office, emphasises how the future office will emerge as a largely integrated system, and draws attention to some of the new commercial expert systems that bear on the office scene. The aim is to convey the flavour of a trend, not to present detailed predictions concerning the office of tomorrow.

INFLUENTIAL FACTORS

There are many important factors which will together influence the shape of tomorrow's electronic office. Any 'broad-brush' listing of such factors should include:

— *Progress in chip technology*
It is research and development in integrated circuits that is largely responsible for the currently available electronic-office systems (word processors, facsimile, mailbox, etc). We may expect the on-going R & D to uncover new system potentials;

— *Progress in other technologies*
There is more to an office system than chips. Technologies focusing on displays and communications (wide-area and local-area networks, fibre optics, etc) are also important. And we may expect person/machine interfaces (eg chip-based speech recognition/synthesis) to develop. Progress in software technology is relevant to the scope of expert systems and other computer facilities;

— *Manufacturer intentions*
It is one thing for technological research to yield a product possibility, quite another for a company to decide to generate a product range. Manufacturers are understandably wary of developing a costly product line which may not be successful in the marketplace. For example, in recent years manufacturers have been reluctant to become 'locked in' to particular network strategies. Such a circumstance naturally affects how quickly particular products are launched on the market. Manufacturer intentions will be broadly conditioned by how they conceive their investment plans, how they rate the state of the national economy, how they perceive international competition, etc;

— *The state of the (inter)national economy*
Any particular company may or may not be adversely affected by a depressed national economy: the firm may

also be swimming in a larger pond. But there are inevitable local factors – availability of skilled labour, prevailing taxation laws, etc – that affect a company's enthusiasm for developing a new product range. The state of the international economy will inevitably affect the commercial buoyancy of national markets, and international factors will influence company attitudes to investment, marketing, etc;

— *The state of company management*
  One possible scenario is where technology has yielded the know-how, where manufacturers have developed the products, where the (inter)national economic factors are encouraging – and yet where a visionless management fails to exploit the possibilities. The awareness and vigour of managements must be rated key factors in influencing the shape of tomorrow's electronic office. We would expect natural selection in the commercial arena to weed out managements (and their companies) lacking the necessary knowledge, drive and imagination.

It is obvious that the evolution of the future electronic office will be mediated by many factors other than the purely technological ones. We may speculate on government education policies as relevant to producing the necessary reserves of skilled labour; on banking policies as relevant to investment (the bank/manufacturer/government relationships in Japan appear to greatly benefit the national economy in that country); on ergonomic research that affects the acceptability (and so commercial success) of new office systems to users, etc. Many factors – technological, commercial, financial, governmental, demographic, etc – are likely to influence the evolution of the future electronic office.

THE INTEGRATED SYSTEM

A central theme in the development of the automated office is the trend towards integrated systems. For many years, observers have noted the convergence of computing and communications, a trend that has encouraged the use of the term 'information technology' to denote a range of different but related concepts and activities. The convergence of various technologies is encouraging the development of integrated office systems. The office workstation will typically be able to handle word processing, data processing, conferencing, mailbox and other functions.

In these circumstances there will be a growing requirement for effective interfaces between various items of office equipment (it is easy to imagine a scenario where it would be useful, for example, to allow a photocopier or facsimile unit to 'talk to' a database facility). Office functions will be increasingly designed as parts of 'total systems', networks of electronic office equipment emerging with particular characteristics shaped by organisational and other influences. Links from items of office equipment to systems running on mainframe computers will become increasingly important. One consequence may be that large companies, specialising in all the relevant kinds of equipment, will experience fewer compatibility problems than will their smaller rivals: companies such as IBM, Xerox, Philips and Siemens may be expected to capture an ever larger market share in competition with smaller suppliers offering a limited range of products.

We have seen that local area networks (LANs) will be increasingly used to connect the various items of electronic office equipment. Such networks are able to connect up compatible and incompatible units to facilitate the development of merged systems and to allow companies to share out their information-technology resources. Ian Ash, the marketing staff manager of General Systems Division UK, has indicated the three elements in IBM office strategy: 'One: stand-alone products; two: systems to extend throughout a single office building; and three: corporate-wide systems. But they must all bolt together'. The emergence of local area networks and other provisions will accelerate the trend towards the integrated electronic-office system. A wide range of computer-based systems will co-operate to provide the office services of the future.

AUTOMATION MILESTONES

We may expect many companies to encounter similar milestones on the route to office automation. As far back as 1979, Price identified five possible milestones in the evolution towards the electronic office:

— document preparation using word processors and phototypesetters exists in an expanding market. It is likely to be, for most organisations, the first step in office automation;

— the second step is likely to be establishing access to data and information bases. This would be done initially by linking to in-house data base facilities. Using facilities outside the organisation would come later;

— the next evolutionary step will probably be intra-company mail;

— this would be followed by access to external data and information bases;

— to be followed finally by an inter-company mail facility.

This scenario charts various steps that a company may take to extend its use of information-technology facilities. Many elements of the electronic office are left out of this framework: individual managements can insert other phases – or develop an entirely different approach – as appropriate to their own organisation and circumstances. The key point is that a comprehensive strategy must be developed – to meet present needs and to allow for expansion and adaptation in the future.

EXPERT (OFFICE) SYSTEMS

Traditional data processing systems store information in various ways, manipulate it (usually in a numerical fashion), and facilitate access by human beings for particular purposes. By contrast, modern expert systems can manipulate stored information (often held in the form of propositional items of expertise) according to the rules of logic, ie expert systems can 'think about' their 'knowledge', reasoning, drawing conclusions, and judging the accuracy of their conclusions. One intention is that expert systems should duplicate the 'thought processes' of human specialists working in a particular discipline. Already, expert systems have achieved impressive results in such fields as medical diagnosis, education, geological prospecting, chemical analysis and electronic circuit fault-finding. It is significant that expert systems are beginning to find their way into the office environment.

For example, the use of expert systems as part of office automation has moved closer for Xionics users. The Isolink telecommunications modules allow users of the Xibus multifunction workstation to access computer systems, databanks, networks, etc without the users needing to know anything about addresses, protocols, procedures or conventions. It is anticipated that Isolink will be able to carry out a range of information processing tasks without reference to the user.

It has been suggested that expert systems will allow managers to improve their performance in the office and other working environments. Alex d'Agapeyeff, principal of Consultants in Information Technology, has pointed out that many companies will use expert systems for recording empirical knowledge ('For example, by recording what their top salesmen really do – not what they say they do – to close orders, then even rudimentary expert systems can marginally improve the performance of the average salesman . . . even a modest improvement in the middle ground means cash in the bank').

A number of commercial expert systems have already been launched in the UK and elsewhere. For example, a number of Nexus-based systems can run on the IBM Personal Computer: available systems include Investment Advisor for investment analysis, and Financial Advisor for business management and the diagnosis of financial problems. Other expert advisory systems are planned for such areas as personnel, loan administration, staff allocation and company tax. Helix, who market the Nexus-based systems, emphasise that the Financial Advisor (for example) is not just a 'black box'. Able to 'think about' such options as overdrafts, long-term borrowing, selling assets, leaseback and credit control, it can advise on cash liquidity problems – and also explain its reasoning. Such advisory facilities are obviously relevant to the efforts of managements seeking to design, implement and operate modern electronic office systems.

Increasing attention is being given in the literature to commercial expert systems for use in a business or office environment (see, for example, Hertz, 1983; and Michaelsen and Michie, 1983). Present funded research (eg in the Alvey Programme) is directly relevant to the provision of expert systems for the office of tomorrow. Such matters will prove of key importance to managers aiming to introduce or extend electronic office systems.

SUMMARY

Many factors will combine to influence the shape of tomorrow's electronic office. These factors variously relate to technology (hardware and software), finance (availability of investment), commerce (the state of national and international markets), government policy (on interest rates, education, etc), and management competence. The electronic office will develop as an increasingly integrated facility, providing a wide range of computer-based services and relying on networking techniques to expand the functional office beyond the confines of a particular geographical site.

New types of computer-based system – in particular, in connection with knowledge processing, expert systems and artificial intelligence (AI) – will have significant effects on the future design and operation of office systems. Speculation has scarcely begun on the effects of, for example, the Japanese fifth-generation programme on office automation, but a route to the cost-effective realisation of integrated office systems can now be charted. Managements should embark upon this route as a necessary condition of company viability in the years ahead, at the same time cultivating the flexibility of attitude that the inevitable technological developments and market trends will demand.

Appendix 1

Bibliography

Bibliography

Beaman I R, *Small Business Computers for First-Time Users,* NCC Publications, 1983

Bird E, *Information Technology in the Office: the Impact on Women's Jobs,* Equal Opportunities Commission, September 1980

Bleazard G B and Price S G, *Managers and the New Technology – the Report of a Multi Client Study,* NCC Publications, 1981

Condon M A, *Office Printers – A Practical Evaluation Guide,* NCC Publications, 1982

Condon M A, *Office Workstations,* NCC Publications, 1982

Damodaran L, Simpson A and Wilson P A, *Designing Systems for People,* NCC Publications, 1980

Firth R J, *Managing Viewdata Systems,* NCC Publications, 1984

Gee K C E, *Local Area Networks,* NCC Publications, 1983

Hertz D B, Artificial Intelligence and the Business Manager, *Computerworld,* 24th October 1983

Michaelsen R and Michie D, Expert Systems in Business, *Datamation,* November 1983, pp 240-246

NCC, *Introducing Computerised Telephone Switchboards (PABXs),* 1982

Peltu M, *Using Computers – A Manager's Guide,* NCC Publications, 1981

Peltu M, *Introducing Computers,* NCC Publications, 1983

Peltu M, *Successful Management of Office Automation,* NCC Publications, 1984

Price S G, *Introducing the Electronic Office,* NCC Publications, 1979

Pritchard J A T and Cole I, *Planning Office Automation – Information Management Systems,* NCC Publications, 1983

Pritchard J A T and Wilson P A, *Planning Office Automation – Electronic Message Systems,* NCC Publications, 1982

Reitz D, The People Side of Word Processing, *CIPS Review,* November/December 1980

Simons G L, *Introducing Word Processing,* NCC Publications, 1981

Wilson P A, *Introducing the Electronic Mailbox,* NCC Publications, 1984

Wilson P A and Pritchard J A T, *Office Technology Benefits,* NCC Publications, 1983

Wood M B, *Introducing Computer Security,* NCC Publications, 1982

Wood M B and Elbra A, *Systems Design for Data Protection,* NCC Publications, 1984

Appendix 2

Glossary

Glossary

Acoustic Coupler — A device used to connect a digital computer-based system to an analogue telephone line through the transmission of sounds (see *modem*)

Backing Store — Computer-based storage, such as magnetic disks or tape, used to hold large amounts of information or software programs that can be read into the computer's main memory when needed for processing

Bandwidth — A measurement of the volume of information that can be carried by a telecommunications link

Batch Processing — A method of data processing common on earlier computer systems. Information is gathered into batches at the computer and then processed sequentially according to priorities allocated in the batch queue

Benchmark — A test to measure the performance of a system

Bug — An error, usually used in relation to software problems. Software products have tended to have more bugs than hardware because techniques of software engineering have been less reliable than equivalent hardware engineering techniques. *Debugging* is the process of seeking out and eliminating bugs

Bureau An organisation which sells a service which processes customers' information on its computers or word processors

Bus The route used to link components within a computer-based system and to provide connections to input and output devices and communications with other systems

Byte A group of eight bits used to represent a character in most computer systems; usually synonymous with 'character' when measuring the capacity of computer storage

Compatibility The ability to move software from one machine to another, to connect a peripheral or storage device to a computer, or to interlink computer-based systems via telecommunications links

Data Synonymous with *information*, mainly used for information that is processed, stored and transmitted in computerised forms

Data Processing Could be applied to any form of information processing. It has become associated primarily with computerised tasks involving large volumes of highly structured information processed in a routine way (such as accounting, payroll processing and other applications evident in the first wave of computing in the 1960s, using large mainframes and batch processing)

Database — Information stored in computerised form. *DataBase Management Systems* (DBMS) are design and software techniques used to organise the database in a structured way so that information in it can be updated, searched, recalled and entered in an efficient, effective and reliable manner

Decision Support System — Computer-based aids which help a decision-maker to reach a conclusion. Decision Support Systems (DSS) include the provision of accurate and timely information to give a realistic basis for arriving at a decision. Computerised *modelling* can be used to investigate possible outcomes if different choices are made. The simplest form of modelling is the *spreadsheet* on microcomputers. This presents a grid of rows and columns of the screen representing a relatively limited number of different factors which are linked by formulae defined by the user

Disks — Magnetically-coated devices, similar to music records, used to store computerised information

Distributed Processing — A system where computer processing capabilities are spread around a number of geographical locations, connected by telecommunications links. A term used in contrast to the centralised DP approach common in the 1960s and 1970s where all computing power was in a central site

EFTS
An abbreviation for *Electronic Funds Transfer System.* This is a service which performs many of the office tasks involved in carrying out financial transactions, including the transfer of money between customer and supplier or between banks

Electronic Filing
A collective term for the storing and retrieving of information using computer-based systems, particularly where the techniques employed mirror the traditional filing structure

Electronic Mail
A computer-based service which enables people to communicate messages, documents and other information using electronic techniques. A number of independent information services could be regarded as part of an electronic mail service such a *facsimile, teletex* and *videotex*. An important feature of a full electronic mail service is the ability to store messages in an area of computer store known as a *mailbox,* and forward them to the recipient(s) when convenient (this service is also known as *store and forward*)

Electronic Typewriter
A means of improving typing productivity without having full word processing capabilities. Electronic typewriters often look similar to electric typewriters but include their own micro chip memory and processing capabilities and, possibly, a small display

Executive Terminal
A workstation designed to optimise managerial, rather than typing or clerical, functions. Likely to have more

	emphasis on electronic mail, numerical calculations and graphical displays than word processing-oriented secretarial workstations
Expert System	A system programmed in a manner to allow system functions that resemble human thought processes. Uses a sequence of detailed logical rules. Human expertise in a specialist area (known as a *domain)* is distilled into these rules in a *knowledge-base*. The system may perform with an equal, or better, success rate than would be obtained by consulting a human expert. A branch of *Artificial Intelligence* (AI) which seeks to create computer systems which behave like people in terms of thinking and communication
Facsimile Transmission (FAX)	A 'telecopier' method of electronically transmitting the copy of a document from one location to another. Fax does not attempt to understand the contents of the document or to put it into computer-readable form; it just transmits the image
Graphics Terminal/ Workstation	For applications that manipulate graphical images, such as in architecture or product design, special graphics workstations and software are needed. A graphical terminal is likely to include a special *high resolution* screen composed of more light dots than a conventional screen so that the image on it is sharper
Handwriting Input	The ability of computer-based systems to understand human writing directly is still relatively limited. Most avail-

able systems can understand a limited range of hand-printed characters (rather than 'joined-up' cursive writing)

Information Retrieval The application of a computer-based system to store information and then to search through it to find what you want

Input The process of getting information into the binary digital code that can be stored and manipulated by computers; for example, using a keyboard, a light pen with a graphics terminal, voice or handprint recognition devices, or a touch-sensitive screen

Intelligent Copier/Printer A system which integrates a number of office capabilities, such as those of a photocopier, facsimile, high-speed printer, word processor and OCR reader

Intelligent Terminal A workstation which has its own computer processing and memory capability so that it can be used independently from any other computing system *(stand alone)* or as a terminal connected to other devices and systems

I/O Abbreviation for Input/Output

Local (Area) Network A means of connecting together a number of computer systems and devices within the same location, typically an office building

Logging On/Off The procedure used to start up a session using an interactive computer-based system is known as *logging on*. Terminating the session is called *logging off*. Logging on should require the input of a special password before allowing access to information

Magnetic Card A form of computerised storage used with early word processing systems. Consists of a card about 8 cm by 18 cm covered in a magnetic coating which can store from about 2 to 10 A4 pages of text. Largely replaced by floppy disks

Magnetic Media Methods of storing computerised digital data according to the direction of magnetisation of tiny magnets in the surface coating of the medium

Main Memory Computer storage, typically held on silicon chips, linked directly to the processor at all times. Used to hold programs and data needed for immediate processing. Provides much higher speeds for transferring information to the processor than backing storage. There are two main types of main memory held on chips, *Read Only Memory (ROM)* and *Random Access Memory (RAM)*. Programs and data in ROM cannot be changed by the user, so is usually used to hold important software used frequently. RAM can be written to and read from. Programs and information can be *loaded* into it, overwriting what was there before

Mainframe Category of computer of the highest capability and power. The term originated because early computers in the 1940s and 50s had a large 'main frame' used to contain the thousands of valves that comprised the central processor

Microcomputer The smallest and cheapest category of computer which can, nevertheless, have the power and capability that once was ascribed to minicomputers and mainframes. They can range from personal computers for home and games use costing less than £50 to sophisticated systems costing up to £9,000 or more, typically about £2,000 to £5,000 for a complete system

Microfilm A compact and relatively low-cost method of storing images of documents for *archival* purposes, where the information is likely to be needed infrequently or as a security back-up to other files

Minicomputer A category of computer between microcomputers and mainframes in terms of capability and price

Modem Devices needed at each end of an analogue telecommunications line to translate between the digital code of a computer-based system and the analogue communications signals. An abbreviation for *mo*dulator/*dem*odulator

Network The method used to interconnect computer-based systems and devices to provide information services covering many locations. A *Local Area Network* (LAN) is based in a self-contained area, such as an office building. A *Wide Area Network* (WAN) covers larger territory, nationally and internationally

OCR (Optical Character Recognition) A means of automatically reading printed characters into a computer system

OEM A company which supplies systems under its own label which incorporate key components or complete computers made by another company; printers, disks and micro or minicomputers are examples of products used by an OEM

OMR (Optical Mark Reading) A method of automatically recognising the position of marks (such as a line in a box on a form) as a method of inputting information directly to a computer

Output The process of getting information out of a computer; for example, by displaying characters and graphs on a screen, words printed on paper, graphs drawn on a plotter, or responding to an enquiry by a computer-controlled voice

PABX or PBX Computerised telephone switchboards capable of providing *enhanced telephony* capabilities and of being used as the basis of a *local area network*

Peripherals Hardware devices other than main memory, processor and workstation or terminal, such as input, output and backing store devices

Personal Computing Computing capabilities used primarily for the personal use of an individual. It can be provided by a stand-alone microcomputer (also known as a *personal*

computer) or a workstation in a network that can also operate as an independent personal computer

Phototypesetting A computer-based system used to produce high-quality printed output as might be required in producing books, magazines and newspapers

Portable Computers Microcomputers which can be carried around. These can be divided into two main categories; *transportables* which are relatively heavy and large; and lightweight and compact true *portables*

Printers A device connected to a computer-based system to produce printed *hard copy* on paper. A printer is a crucial element in the electronic office

Programming Languages Languages with clearly defined rules and grammars used to write software

Software The programs which automatically control the operations of computer hardware. There are two main categories of software; *systems software* that manages the machine's resources, and *applications software* that handles the user's tasks directly

Tape Magnetically coated devices, similar to music tapes and cassettes, used to store computerised data. Tiny magnets in the surface of the tape are used to represent bits, which can be set or interpreted by a read/write head in the tape deck or cassette recorder

Telecommunications The transmission of information across distances using electronic and other methods not involving physical transportation. Transmission can be either *digital,* consisting of discrete bits of data, or *analogue,* represented by a continually varying representation such as sound waves

Teleconferencing A service which enables meetings to be held with participants at various locations communicating by telecommunications networks. Participants usually have to gather in special rooms with suitable equipment

Teletex Text communications service based on standards set by the CCITT, an international telecommunications body, interlinking word processors, electronic typewriters, special workstations and special teletex workstations

Teletext Text and graphic information broadcast from a central point to displays adapted to receive the encoded information. Ordinary TV sets can have a TV teletext adaptor added to them

Terminal A workstation or other device which has a direct telecommunications link into a computer-based information network

Tickler A software capability that enables the user to be reminded to perform some task. The user sets up the 'tickler' file with the appropriate deadline and action that needs to be taken. A warning is then given when the deadline comes near

Video Disks and Tape Methods of storing images in digital form, similar to entertainment video disks and domestic video cassette recorders

Videotex An information service combining text and graphics, containing a database, usually divided into *pages,* which can be linked to a network of terminals, including adapted TV sets

Viewdata Text and graphic database and information service provided via terminals that can communicate with the system's computers and with other terminals. Viewdata was originally developed by British Telecom as the Prestel public viewdata service using domestic TV sets as terminals

Voice Processing Methods of handling voice data in a computer-based system. *Voice recognition* is a form of computer input that directly interprets human voices. *Voice response* is computer-controlled speech output which can be created by generating synthesised voices made artificially from computer-generated sounds

Word Processing A computer-based system for handling textual information (also known as *text processing*). Word processing can take place on a purpose-built word processor containing special software, by using word processing software on a general-purpose computer, or via a terminal with access to word processing software

Workstation Any device from which work can take place using computer-based aids. *Multifunction workstations* can carry out a variety of tasks, such as word processing, numerical calculations, electronic mail and voicegrams. Other workstations are designed for special functions, such as word processing, electronic mail, graphics or information retrieval

Appendix 3
Sample Tender Document

Sample Tender Document

THE ABC COMPANY LIMITED
OFFICE SYSTEM REQUIREMENTS

Prepared by:
A N Other
The ABC Company Limited
Street Lane
East Chester
E1 8XY
Tel: 071-623 2345

Date:

CONTENTS

1 Introduction
2 Office System Applications
3 Suppliers' Proposals
Appendices

1 INTRODUCTION

1.1 Background

Following an Office System Feasibility Study, the ABC Company Limited (subsequently referred to as ABC) are now inviting office system manufacturers/software houses to submit proposals to fulfil the office system requirements outlined in this report.

1.2 The Company

ABC, founded over 70 years ago, is a wholesaler and distributor of a selected range of motor-car accessories. Current turnover is approximately £3 million and the company expects to increase the amount of business by three times in the next five years.

The company deals almost exclusively with account customers only. All items carried are described in the ABC catalogue and are identified by a list number, which the customer uses to order an item. Certain list items are composed of more than one assembly item which are uniquely identified in the stores.

1.3 Purpose of the Report

This report is the document upon which suppliers are asked to base their proposals.

General information about the proposed computer-based office systems is provided, so that the type of equipment configuration can be determined.

Outline descriptions of each unit of the proposed system are given so that a quotation may be prepared for the detail system design work, programming, testing and implementation.

Section 3 shows the structure which the proposals should take and the information they should contain.

Although both Section 2 and the Appendices are broken down into a number of units or subsystems, this is for convenience of specification only and the solution to be aimed for is a fully integrated system.

Initially, ABC wishes to minimise its own staff involvement in systems design and programming.

It is intended that a single contract is placed for both equipment and software. It is, however, appreciated that an office system manufacturer and a separate software house may be required. Should this be the case, the leading organisation should prepare the proposal giving details of any other organisations who may be employed as sub-contractors for providing the equipment, systems design or programming. The leading organisation should ideally take contract responsibility for the system as a whole.

2 OFFICE SYSTEM APPLICATIONS

2.1 General

The following paragraphs describe briefly the applications which it is intended to implement, more detailed information on volumes and field sizes is contained in the Appendices as a guide to file sizing. Implementation of the office systems is to be phased.

.
.
.

details of required office systems

.
.
.
.

3 SUPPLIERS' PROPOSALS

The proposal should contain the following information and relate closely to the headings specified. A standard approach is necessary in order to simplify evaluation and ensure that each proposal meets the requirements specified.

3.1 Introduction and Summary of Proposal

A summary page should be given listing the equipment proposed and overall costs for the hardware, maintenance, applications software and system software.

3.2 Recommended Equipment

The following information should be given:

— details of the operating characteristics of each item of equipment showing costs for purchase and rental;

— the capability for expanding the configuration;

— the mode of degradation in the event of power failure.

3.3 Software

This should give details of all items of software with costs where relevant, which the company would be likely to use or which could be required to operate the equipment and systems. It should give the following information where applicable.

3.3.1 Overall Picture

Concise description of proposed systems showing, in particular, any system integration.

3.3.2 Applications Packages

— basic concepts;

— policy on amendments/additions;

— ownership;

— statutory changes, program maintenance;

— guarantee period.

3.3.3 Tailored Systems

— precise definition of services offered;

— software responsibility, sub-contractor policy and overall responsibility;

— calibre and experience of staff; for software houses, the period for which the software house has been in existence and number of staff;

— program maintenance responsibility;

- error notification and response procedure;
- guarantee or acceptance period.

3.3.4 System Software

- operating system functions;
- age of particular operating system and number of users of version specified;
- maintenance and enhancement policy;
- utilities;
- systems and programming aids;
- recommended language;
- other possible languages;
- report generation software;
- screen format generation method;
- ad hoc enquiry facilities;
- communications.

3.4 Support

The support which can be provided should be itemised including costs where relevant. The following information should be given:

- account responsibility;
- training facilities and costs (systems/installation and programming);
- time required for scheduled maintenance;
- response time to called-out maintenance;
- the location of staff for maintenance cover and the cost of called-out and on-call services;
- the extent and nature of back-up services in the event of prolonged machine failure;
- any other support available.

3.5 General

- — installation requirements and related costs;
- — the nature of systems documentation provided;
- — recommended implementation procedure and plan;
- — delivery date;
- — insurance and delivery costs;
- — costs of ancillary supplies (disk packs, paper, etc);
- — names and addresses of companies and persons for organisations operating similar applications and using similar equipment to whom reference may be made;
- — number of UK users of proposed model;
- — number of UK users of the standard software;
- — list of local users;
- — copies of suppliers' standard contract terms covering rental, purchase, maintenance and supply of software or services.

APPENDICES

•
•
•
•
•

Appendix 4
Sample Letter to Suppliers

Sample Letter to Suppliers

Dear........................

I enclose an outline Statement of Office System Requirements presented on behalf of the ABC Company Limited.

I should be grateful if you would confirm that you are interested in submitting a proposal and that it will be available by the

....

Any queries regarding the content of the report and the requirements of the ABC Company Limited should be directed to myself.

...............copies of the proposal are required by the above date.

I look forward to hearing from you further.

Yours ....

Enc.

Appendix 5
Cost Elements

Cost Elements

CAPITAL COSTS

Computer Equipment	—	all items including CPU, VDUs, printer, storage devices, etc;
Software	—	all system software including operating system, utilities, language compiler; applications packages, tailor-written programs;
Data Communications Equipment	—	eg modems, acoustic couplers, line drivers;
Site Preparation	—	provision of clean power supply; air conditioning; flooring, partitioning;
Supplies	—	desks for VDUs etc; data storage cabinets; fireproof storage media, eg disks or diskettes; stationery, both standard and special;
Implementation	—	equipment delivery; documentation; training for management, operators and possibly programming;
Expansion	—	extra hardware units and upgrades; additional applications software;

File Creation	—	data collection and verification; data input and verification.

RUNNING COSTS

Staff	—	any new staff specifically employed for the system; any marginal salaries due to upgrading existing staff for the system; any staff employed to replace operators, etc, in their old jobs;
Supplies	—	consumables such as stationery, printer ribbons, etc;
Maintenance	—	annual contract for maintenance and support of all hardware items; either support contract or contingency for software maintenance, support and updates;

OTHER COSTS

	—	hardware insurance; additional training.

NOTE:

Some capital costs especially for hardware and software may actually be annual costs if, for example, the hardware is rented or leased, and if the software is used under licence.

Appendix 6
Checklists

CHECKLIST A	:	SYSTEM DESIGN AND OPERATION
CHECKLIST B	:	DESIGN, IMPLEMENTATION AND OPERATION
CHECKLIST C	:	DIAGNOSIS AND AWARENESS
CHECKLIST D	:	HUMAN ASPECTS CRITERIA (VDUs, KEYBOARDS)

Note: Checklists can be represented in many forms for many purposes. Taken together, the checklists in this appendix cover the various aspects of systems diagnosis, design, implementation and operation. The checklists, to some extent overlapping, can be adapted to specific management and design requirements.
(Sources: Checklists A and D – Damodaran et al, 1980;
Checklist B Peltu, 1981;
Checklist C – Pritchard and Wilson, 1982)

Checklist A
System Design and Operation

1.1 SETTING UP LINES OF COMMUNICATION

1 List the people in the following categories who will need to be kept informed of the development and its progress:

- users;
- management;
- unions;
- user associates.

2 Identify the information needs of each of the above categories of people.

3 Decide on the methods by which the information will be disseminated and feedback can be obtained.

4 Make one or more individuals responsible for information dissemination over the life of the development.

1.2 PLANNING THE DESIGN STRATEGY

1 Become acquainted with the spectrum of design strategies that are available.

2 Identify:

- the management who will control the development;
- the potential users of the proposed system;
- the designers of the proposed system.

3 Interview users, managers and designers to establish what experience they have had of system developments, systems design work and design strategies. Use the interviews as an opportunity to get familiar with the work environment and to establish a rapport with the users.

4 Decide on a particular strategy to obtain constructive user contributions to the design, bearing in mind the benefits and problems involved.

5 Plan any necessary training to allow users, designers and management to follow the chosen strategy.

6 Inform all parties of the strategy that is to be employed and ensure that they each understand the roles that they will have to play.

1.3 PLANNING FOR DISRUPTION

1 Identify any disruptions that are likely to occur in the following categories:

- management control activities;
- user contributions to systems design;
- physical changes to the workplace;
- installation of equipment and furniture;
- training the users to operate the new system;
- system testing, parallel running and system initiation.

2 Plan to cater for the disruptions identified in 1 above.

3 Ensure that all potential disruptions are understood and catered for.

1.4 WORK SITUATION APPRAISAL

1 Familiarise yourself with the user's work environment, ie gain an appreciation of the nature of the work, the pressures, priorities, skills, values, and attitudes of the people.

2 Identify the specific human factors information to be collected.

3 Review the available data collection techniques and select those likely to be appropriate and acceptable.

4 Identify the routes to follow through the organisational structure in the detailed analysis phase.

2.1 USER IDENTIFICATION

1 Identify potential users by identifying the name and job titles of the users

of the existing system.

2 Develop a 'map' of the user population. For example, it will be helpful to classify future users in terms of primary, secondary or tertiary users to avoid focussing solely upon the main users at the expense of the others.

3 Use the completed 'map' as a directory of potential users when user needs and characteristics have to be established at various points in the development.

2.2 TASK DEMANDS

1 Identify the range and functions undertaken by the users with particular emphasis on the variations in the work.

2 Review techniques for task analysis and select appropriate techniques.

3 Conduct detailed analyses of user tasks with emphasis on task differences.

2.3 WORK ROLE ANALYSIS

1 Gain an understanding of the user's perception of his work, its purpose, value and meaning for *him*.

2 Gain the user's views on the rewards and costs of the job.

3 Identify the most important aspects of the job for the user.

2.4 USER READINESS

Assess the user's receptivity/state of readiness for the proposed system. For example:

1 Assess user's previous experience of computing.

2 Examine user attitudes to computer use.

3 Assess user knowledge of the proposed computer application.

4 Assess user's willingness to cooperate in system development.

3.1 THE MECHANICS OF THE SELECTION PROCESS

1 Establish a small committee which will have responsibility for items 2 to 5 on this checklist.

2 Draw up lists of:

- functional criteria;
- general criteria;
- human aspects criteria.

3 Identify essential criteria.

4 Assign weighting to each of the identified criteria.

5 Identify the factors on which each criterion is to be judged; then assign a scoring system to the factors.

6 Establish which companies manufacture the equipment you require and contact them. Inspect their models and discuss your requirements. Eliminate those companies whose equipment does not meet the essential criteria. Draw up a shortlist of equipment for final evaluation.

7 Evaluate each model on the shortlist by the weighted ranking method and select the equipment on the basis of the results.

4.1 JOB DESIGN

1 Identify all tasks, old and new, which will undergo change as a result of the implementation of the computer system.

2 Guard against unchallenged assumptions about the manner in which tasks are combined to produce jobs. In particular guard against too much task specialisation, especially with respect to data input.

3 Generate alternative ways of allocating tasks to jobs by considering:

- approaches such as job rotation, job enrichment and autonomous work groups;
- allocations by function, product or customer;
- the implications of the technical system.

4 Provide opportunities for user groups to explore the consequences of the alternatives by embodying the ideas in trial systems or simulations, and assist the process of determining a job design solution.

5 Retain flexibility in the implemented system in order that user groups can continue to adjust the allocation of tasks to jobs to suit emerging demands.

4.2 ACTIVITY FEEDBACK

1 Identify the individuals for whom activity feedback is to be provided. Establish what tasks they will be performing.

2 Decide the activity feedback information to be provided for each individual. Discuss the choice of information with the individual concerned and ensure that he feels the activities chosen are significant parts of his job.

3 Consider how the activity feedback information selected in 2 relates to existing management control information.

4 Design the details of the activity feedback information such as:

- how the information will be collected;
- how it is to be presented;
- the timing of the information.

5 Ensure that the individual for whom the information is to be provided knows:

- how the information will be collected;
- how the information is analysed;
- how to interpret the information when he receives it.

4.3 DIALOGUE DESIGN

1 Obtain the following information:

- details of the activities to be undertaken by the users;
- roles that the dialogue users play in the working situation;
- technical jargon, abbreviations, codes or cultural linguistic differences that are specific to the users;
- details of the equipment to be used to conduct the dialogue;
- details of dialogue design standards already existing in the organisation, and details of any dialogues which the users already encounter on a regular basis.

2 Remember the following general principles:

- plan to put a lot of effort into dialogue design;
- apply dialogue design standards;
- design a flexible dialogue;
- ensure that the dialogue is logical and easy to use from the user's point of view.

3 Identify the data elements that are likely to occur in the dialogue and define their characteristics.

4 Select a dialogue style.

5 Break down the dialogue into discrete blocks produced alternately by the computer system then by the user.

6 Design the dialogue for each block, for users of intermediate capability.

7 Design validation procedures and error messages.

8 Design the screen formats.

9 Test and modify the design.

10 Design flexibility into the dialogue to accommodate new users, very inexperienced users, very sophisticated users, and/or other categories of user.

4.4 DESIGNING PROCEDURES FOR FAULTS AND BREAKDOWNS

1 Design procedures for rectifying:

- system faults;
- system breakdowns;
- terminal faults;
- terminal breakdowns.

2 Design alternative procedures for business operations when:

- a system breakdown occurs;
- a terminal breakdown occurs.

3 Design procedures such that users can take positive action when they are confronted by a fault or breakdown. Ensure that users are trained in these procedures.

4 Design procedures for the recording of faults and breakdowns.

5 Ask users for their constructive criticism of all of the above procedures which will affect them. Where possible ask them to test the procedures. Make any necessary modifications.

5.1 COLLECTION OF ENVIRONMENTAL DATA

1 Get agreement from user management to carry out a preliminary survey of the workplace. Ensure that user management informs the occupants of the workplace about the survey and its purpose before it takes place.

2 Measure the workplace. Use the results to produce a small-scale plan of the workplace which is suitable for photocopying.

3 Return to the workplace with 6 copies of the small-scale plan. Mark details in rough of each of the following on separate plans:

- layout of equipment and furniture;
- access requirements;
- floor characteristics;

- position of air conditioning or heating equipment;
- position and type of lighting;
- position and type of plug sockets and cable runs.

4 Check noise and vibration characteristics of all equipment.

5 After all the data has been collected, clearly record and label it on appropriate plans so that it is easy to read and understand.

5.2 WORKSTATION DESIGN

1 Investigate:

- the equipment that is to be used;
- the tasks that are to be undertaken in conjunction with the equipment;
- any other tasks that are to be undertaken at the workstation;
- the human beings who will be using the workstation;
- the layout of the workplace in which the workstation is to be located;
- the workstations currently in use.

2 Decide on the following basic design parameters:

- the user population;
- sitting or standing workstations;
- heights;
- viewing distances;
- eye and head movement;
- document holder requirements.

3 Establish worksurface dimensions and layout by considering the following criteria:

- arm reach;
- size of equipment;
- handedness;
- task requirements;
- use of documents or manuals;
- display of notices;
- office space constraints.

4 Determine the characteristics of the chair that will be required. Inspect chairs available on the market until two or three options that fit the bill are found. Determine the characteristics of the footrest that will be required.

5 Establish the details of the workstation design by considering the following factors:

- leg space;
- storage space;
- modularity;
- ease of cleaning;
- cabling;
- materials;
- weight;
- mobility;
- floor covering;
- waste disposal facilities;
- safety.

6 Check the design to ensure that the following activities have been catered for:

- the setting and monitoring of the terminal controls;
- use of the equipment;
- maintenance of the equipment;
- the installation and replacement of the equipment.

7 Produce a prototype workstation. Draw up a list of factors to be evaluated and the procedures for evaluating each factor. Evaluate the workstation under as realistic conditions as possible. Ensure that any muscular strain, visual fatigue or other health and safety hazards are identified. Modify the design and evaluate again.

Continue this process until a satisfactory design is obtained.

5.3 LIGHTING

1 Ask the occupants if they think the lighting in the room is satisfactory.

2 Obtain the information about activities established at the job design stage.

3 Consult the IES code to establish:

- recommended illuminance levels;
- a suitable colour appearance for the lamps for the particular activities to be undertaken in the room.

4 Measure existing luminance levels in the room, and determine the colour appearance of the existing lamps. Match these against the requirements established in 3 above. Decide if new lamps are required. If so, ensure that the lamps that are purchased will provide the required illuminance levels when fitted with the appropriate glare shielding.

5 Establish if a glare problem exists in the room. If so, fit appropriate glare shielding. Ensure that any glare shielding fitted to the lamps does not reduce illuminance below the required levels.

6 Ensure that the reflection levels in the room are satisfactory.

7 If any changes have been made as a result of following the above steps, ask the occupants if they think the lighting in the room is now satisfactory.

8 Record the new position of lamps and/or local lighting on the scale plan of the room.

5.4 ROOM CLIMATE

1 Investigate the following physical characteristics of the room:

- the heating system;
- doors;
- windows;
- radiant temperatures.

Also record the room temperature and outside temperature (morning, midday and evening) for a period of two weeks.

2 If any equipment is to be installed in the room, inspect it in operation before installation and gauge the amount of heat output and the amount of air flow generated.

3 Spend some time in the room. Ask the occupants:

- if the room sometimes gets too hot;
- if the room sometimes gets too cold;
- if the room sometimes gets too stuffy;
- if there are draughts – if so where do they come from;

– what they think of the heating and ventilation conditions generally.

4 Reach a decision with the appropriate management about what changes, if any, are required to the room climate. This decision will probably take the form of either doing nothing, or instituting minor changes (such as installing window blinds or getting the heating system overhauled) or of calling in a heating engineer to undertake more detailed work.

5 If any work is carried out on the room climate, visit the room about 1 month afterwards and ask the occupants what they think of the new conditions. Make three other similar visits – each visit corresponding with the middle of a season (room climates need to be checked in all of the seasons of the year). Decide if further work is required. If so, carry it out and continue the checks until a satisfactory room climate is obtained.

5.5 NOISE

1 Spend some time in the room. List the noises that can be heard and identify their sources. Ask the occupants' views on noise levels in the office.

2 If equipment is to be used in the office, listen to it in use before it is installed. Consider the noise level figures quoted by the manufacturer. Decide if it worth trying to purchase a silencer or reduce the noise output in some other way. If so, pursue that course of action.

3 Assess the general level of noise in the environment. If it is too high, consider the following possible remedies:

– increasing the sound absorption capabilities of the room with such things as carpeting and sound absorption room dividers;

– reducing the levels of other sources of noise in the room such as telephone bells;

– calling in a noise engineer.

4 If noise reduction work has been undertaken, or if new equipment has been installed in the office, review the noise conditions after a month or so. Ask the occupants of the room what they think of the noise conditions in which they work. Decide if further work is required. If so, carry it out and continue to review the office noise conditions until they are satisfactory.

5.6 ROOM LAYOUT

1 Establish:

– if any work groups are to be kept physically together;

– if new work groups are to be created;

- if some people do not want to be located close to others (eg, smokers and non-smokers);
- if some people need to be located close to others to facilitate the work to be done.

2 Determine what sort of office is to be set up:

- open plan;
- individual offices;
- mixture of both.

If either of the latter two are decided upon, establish which personnel are to be located in which type of office.

3 Establish what space standards, if any, are to apply. List all furniture and equipment that are to be included in each room. Calculate the space availability. If it is unsatisfactory, change the numbers of people or contents of the room until the space availability is satisfactory.

4 Use the general room information obtained during the collection of environmental data to draw a large-scale plan of the room. Ensure that doors, windows and power and communication points are all marked on the plan. Or, prepare a link diagram, using information on the connections between people and people, and people and equipment.

5 Shade in those areas of the room which are unsuitable for siting workstations. Identify such positions by considering the following constraints on the workstation positioning:

- direction of light;
- avoiding glare;
- avoiding reflections on VDU screens;
- avoiding positions subject to excessive heat or draughts;
- avoiding positions subject to excessive noise;
- position and size of major passageways;
- door access requirements;
- window access requirements;
- equipment maintenance access requirements;
- fire-fighting equipment access requirements.

6 Make scale cutouts to represent the furniture, workstations and equipment to be fitted into the room. Try out various arrangements of the cut-

outs on the plan. When possible alternatives are established, evaluate them against:

- work group requirements;
- constraints on the positioning of workstations;
- health and safety criteria for the routeing of cabling.

7 Present the possible layouts to the potential occupants and ask them for their comments.

8 Make a final decision about the layout that is to be implemented, bearing in mind the views of the occupants. Plan to implement the room layout taking care to cater for the disruption caused to the occupants.

Implement the layout.

9 Establish identifiers for each piece of equipment and label the equipment accordingly. Make an entry for each terminal on a Terminal Identification Record.

10 Review the layout after it has been in operation for 2 or 3 months and make any necessary changes.

5.7 MAINTENANCE CHECKS

Equipment

1 Is the screen clean? If not, clean it.

2 Is the image clear? If not, record details.

3 Is the image stable? If not, record details.

4 Are the keys clean? If not, clean them.

5 Are the top, sides and underneath the equipment clean? If not, clean them.

6 Any complaints about response times? If so, record details.

Workstation

7 Is there sufficient space for papers and reference documents on the workstation surface? If not, record details.

8 Is the workstation surface tidy? If not, tidy it.

9 Are the workstation drawers tidy? If not, tidy them.

10 Can the drawers be opened easily? If not, record details.

11 Is the general condition of the workstation desk satisfactory? If not,

record details.

12 Is the general condition of the workstation chair satisfactory? If not, record details.

13 Does the chair adjusting mechanism work satisfactorily? If not, record details.

14 Do all the users know how to adjust the chair? If not, record that training is required.

15 Are the wastepaper facilities adequate? If not, record details.

16 Is the condition of the notices displayed on the workstation satisfactory? If not, record details.

17 Is it easy to get in and out of the workstation? If not, record details.

18 Is the condition of the footstool satisfactory? If not, record details.

19 Is any wiring cluttering up the workstation? If so, record details.

Lighting

20 Are any of the room lights not working? If so, record details.

21 Are any of the fluorescent room lights flickering? If so, record details.

22 Are there any sources of glare in the room? If so, record details.

23 Do any of the lights, lampshades or diffusers need cleaning? If so, record details.

24 Are the window blinds easy to adjust? If not, record details.

25 Do the window blinds effectively shield the occupants from the sun's rays and from glare? If not, record details.

Room Climate

26 Is the room sometimes too hot or too cold? If so, record details.

27 Are there any draughts in the room? If so, record details.

28 Is the room affected by dust, fumes or smells? If so, record details.

29 Does the room get stuffy? If so, record details.

Noise

30 Is the room too noisy? If so, record details.

Room Layout

31 Is there adequate unobstructed passageway space in the room? If not,

record details.

32 Does any wiring cross any passageways? If so, remove it.

33 Does any wiring obstruct the feet in the course of using the workstation? If so, record details.

34 Is any of the wiring severed or frayed? If so, take any dangerous wiring out of service and record details.

35 Are there any broken or loose plugs or sockets? If so, take any dangerous plugs or sockets out of service and record details.

5.8 UPKEEP

1 Allocate responsibility for the upkeep of equipment, workstations and the workplace to a specific job title.

2 Identify the checks that must be made, or work that must be carried out, to undertake the planned maintenance activity.

3 Specify the time period between planned maintenance inspections for each check.

4 Establish who is to carry out each maintenance check.

5 Establish the procedure for obtaining the view of the users when planned maintenance is carried out.

6 Establish procedures for undertaking each item of planned maintenance and for reporting on planned or breakdown maintenance activity.

7 Draw up a planned maintenance schedule and record it on a suitable planning chart.

8 Establish procedures for monitoring maintenance records to identify any inherent weaknesses or likely future breakdowns.

6.1 PLANNING USER SUPPORT

1 Appoint one person to be responsible for the design and implementation of the user support network.

2 Establish:

- the users of the new system and the relevant skills or knowledge they possess;
- what new tasks the users will have to perform;
- details of existing human support mechanisms;
- details of existing documentary support;

- details of existing change mechanisms.

3 Plan human and documentary support.

4 Plan change mechanisms.

5 Plan the training programme.

6 Review the support network and make modifications where necessary.

7 Implement the support network.

6.2 TRAINING

1 Identify all the people who will be affected by the new system.

2 Establish the training needs of each individual.

3 Determine teaching objectives for each need established.

4 Identify appropriate training modules.

5 Determine what techniques are to be used to carry out each module. Plan the content of each module.

6 Arrange times, dates, places and people for the programme to be actually carried out. Ensure that:

- the trainees understand why they are undertaking the training programme;
- user management has set aside appropriate time for staff to undertake the programme.

7 Carry out the programme, monitor and control it to ensure that the objectives are achieved within time and resource constraints.

8 Review the training programme. Identify:

- any improvements that could have been made;
- any outstanding or future training requirements.

Feed this information into the next training programme.

6.3 DOCUMENTARY SUPPORT

1 If terminal equipment is being installed obtain the following documents from the terminal manufacturers:

- installation instructions;
- maintenance instructions;
- the technical specification of the terminal;

– the terminal operation guide.

Pass the installation and maintenance instructions, and the technical specification to the job title responsible for maintenance activities.

2 Establish what documentation currently exists in the organisation. Consider the job of each user in turn. Decide whether he will require a user's guide to equipment operation, a user's guide to the system or any other documentation.

3 Decide which of the following methods of providing documentary support is to be used for the user documentation:

– paper-based manual;

– paper-based circulars;

– expert system;

– help facilities;

– computer-based documentation.

4 Use your own internal standards for producing the documentation. If no standards exist, decide on the format, style, presentation and method of update to be used. Get the user to assist in producing the documentation or, failing this, to vet it as it is produced.

5 Get some of the potential users to try out the documentation and note their reactions and comments. Make any necessary modifications and try it out on the users again. Continue this process until the documentation is satisfactory.

6 Set up procedures that will enable users to suggest changes to the documentation. Display these procedures in a prominent position, either in the documentation or on the documentation workstation.

7 Establish which job title has responsibility for making amendments to the documentation.

8 For paper-based documentation, establish which job title has responsibility for the physical update of the manuals.

6.4 HUMAN SUPPORT

1 Read about the various types of human support mechanisms in the relevant section of this book.

2 Establish what human support mechanisms already exist within the organisation. Determine how successful they are. Consider the activities each user will be carrying out. Decide what human support, if any, he should be provided with.

3 Set up the human support mechanisms required. Ensure that the support person:

- fully understands the job he has to do;
- is capable of carrying out the job;
- is formally recognised to be carrying out the support function.

4 Ensure that the users concerned know what human support is available for their use.

5 Review the human support mechanisms after their first 6 months of use. Seek user opinion. Ask the people providing the support how well they think the mechanism is working. Make any appropriate modifications.

6.5 CHANGE MECHANISMS

1 Read about the various types of change mechanisms in the text of this section.

2 Determine what change mechanisms currently exist and how successful they are.

3 Decide what change mechanisms will be most appropriate for each user.

4 Set up the change mechanisms and ensure that appropriate feedback mechanisms are also instituted.

5 Review the change mechanisms after their first six months of use. Seek user opinion. Make any appropriate modifications.

6.6 CONTINUING USER SUPPORT

1 Ensure that an appropriate amount of training is given to users when changes are made to the system.

2 Establish who the local experts are. Find out just what knowledge and abilities they have that make them local experts. If they are about to leave the organisation find a means of making that knowledge and capability easily available to users until a new local expert emerges.

3 Draw up a training programme for new starters. Ensure that the programme includes the following:

- training in the broad aspects of the system and its interfaces;
- hardware acquaintance training;
- skills and detailed system training;
- training in how to adjust workstation chair and workstation equipment;

- training in what documentary and human support is available;
- training in what change mechanisms are available.

Apply the organisation's policy on eyesight to the new starter. Be aware that there is likely to be a continuing training need. Meet the needs by the most appropriate training mechanisms.

Checklist B
Design, Implementation and Operation

Question/phase	Checklist
WHY? Corporate objectives	— Establish management structure and procedures to be used for information management — Set up task forces or appoint consultants to undertake feasibility studies (in consultation with relevant managers) — Identify: — Products/services to be enhanced (known as *value-added analysis)* — Activities requiring productivity improvement — Main costs which need to be cut — Pre-emptive actions or responses to meet competitive challenge — Major social influences shaping organisational objectives and behaviour — External legislative/political/economic requirements
WHERE?	— Gather facts on existing systems
Feasibility studies	— Evaluate areas of strength/weakness in existing methods

Question/phase	**Checklist**
	— Create management awareness of technical potential
	— Ensure technical staff is aware of real organisational needs
	— Determine procedure for consultation with staff on technological change and system design (this might result in a formal technology agreement)
	— Set criteria for presentation of results of feasibility studies, systems design proposals and operational evaluation, insisting on lucid, plain language descriptions and systematic documentation procedures
	— Prepare cost/benefit analysis on alternative approaches to meeting corporate objectives which take into account *total* systems cost
	— Define likely conversion requirements in moving to new systems
	— Identify most suitable approach(es)
	— Develop overall systems strategy
	— Identify applications to be computerised in a phased implementation plan
WHAT? Systems specification	— Define procedures for managing and monitoring the design process
	— Establish design project teams which balance technical specialists with the management and staff who will use the system
	— Prepare detailed applications specifications
	— Ensure high priority is given to human factors, both in staff needs and customer/client requirements

Question/phase	Checklist
	— Identify long-term needs for expanding the system's capabilities
	— Define corporate technical standards aimed at simplifying and cutting the costs of systems growth (for example, standards for programming languages, databases and communications networks)
	— Specify likely conversion costs, irrespective of the computer system chosen
	— Set high-performance standards based on real needs and not limited by preconceived and possibly inaccurate expectations of the capabilities of computers
	— Identify staff resources and training needs both for technical staff and users
	— Define security requirements
	— Ensure the system incorporates an 'audit trail' for use by external or internal auditors
	— Consider requirements relating to the protection of personal data
	— Establish implementation schedules giving weight to critical corporate needs but taking into account realistic systems development timescales
HOW?	— Collate information on potential suppliers
Evaluation of suppliers	— Prepare cost estimates for different systems approaches and suppliers
	— Identify one or a number of feasible approaches and potential suppliers who could be approached
	— Prepare a detailed invitation to tender which summarises the systems specification

Question/phase	**Checklist**
	— Evaluate responses to tender and prepare a short-list which takes into account, in addition to technical suitability: — proof of supplier's commercial viability — number and range of supplier's customers — proximity of existing users — quantity, quality and location of support staff — documentation and training support
	— Choose supplier(s) and products, services and financing method (buy, rent or lease)
	— Draw up an agreed contract with supplier(s) which explicitly covers a wide range of possible eventualities
	— Insure the system
WHEN? Implementation	— Establish management structure to coordinate implementation process
	— Prepare detailed implementation schedule
	— Identify key tasks that need to be undertaken with appropriate allocation of time and staff resources
	— Develop detailed staff recruitment and training plans for specialist computer staff and in user departments
	— Create strategy covering phased progress of pilot projects, testing and eventual 'live' operation
	— Prepare site for computer installation
	— Define detailed programming and systems specifications to pre-defined standards which promote good software engineering

Question/phase	Checklist
	— Specify level of technical support needed to maintain adequate service and establish procedures to ensure such support is provided
	— Ensure effective procedures are adopted to monitor and evaluate systems development and implementation
	— Establish suitable back-up facilities
	— Avoid trying to do too much too quickly
AND THEN . . . ?	— Establish management strategy
Post-implementation	— Compare performance and cost with original objectives in continuous reviews
	— Identify benefits achieved and problems created
	— Continuously monitor performance of suppliers
	— Evaluate performance with a view to identifying benefits that could be obtained from enhancements to existing applications and opportunities for new areas of computerisation
	— Ensure there is a systematic procedure for monitoring new technological developments to see how they can be integrated with or replace the existing system
	— Give high management priority to establishing and monitoring the status and effectiveness of the computer services and/or computer operations department (if the workload is sufficient to justify such departments)
	— Continuously check that the documentation relates to the actual system and that documentation does not deteriorate over a period of time

Question/phase	Checklist
	— Make a computer audit as routine a part of management appraisal as a financial audit

Checklist C
Diagnosis and Awareness

1 *The Business*

1.1 Who owns the business and what do they want out of it?

1.2 What are the prime activities that the business undertakes and where do the major costs lie?

1.3 What is contained in the organisation's corporate plan (if it has one)?

1.4 What informal policies does the organisation have?

1.5 What do individual senior managers perceive the organisation's objectives to be? What changes do they perceive in the future?

2 *Organisation Structure*

2.1 Functional activity model.

2.2 Organisation charts showing every member of staff.

2.3 Geographical locations.

2.4 Comparison of the three types of organisation identified:

— functional activities;

— departmental organisation;

— geographical location.

2.5 What are the shortfalls of the current organisation?

2.6 How might the organisation change in the future?

2.7 How must the organisation of service departments change to take advantage of the convergence of DP, telecommunications and text handling?

2.8 How do the geographical locations of various functions and departments help or hinder the organisation's operations, and how can electronic message systems help to rectify the problem?

2.9 How can one ensure that all locations will take up the new technology?

3 *The Staff*

3.1 Total salary bill; turnover per head; profit per head.

3.2 Headcount and salary bill of full-time and part-time staff.

3.3 Headcount and salary bill of:

— senior management;

— middle management;

— professional and technical staff;

— clerical, secretarial and administrative staff;

— manual staff.

3.4 Headcount and salary bill of each department.

3.5 Average length of service and average age of:

— total workforce;

— senior management;

— middle management;

— professional and technical staff;

— clerical, secretarial and administrative staff;

— manual staff.

3.6 Headcount per department of those with:

— experience of using a VDU (either before or after joining the organisation);

— experience or knowledge of any advanced office systems.

3.7 Membership of trade unions/staff association by:

— department;

— geographical location.

3.8 Headcount and salary bill per geographical location.

3.9 Past values and future trends of all of the above.

3.10 The contents of any policy statements issued by the central union organisation.

3.11 The contents of any policy statements issued by any or all of the branches within your organisation.

3.12 Recent reactions of unions or staff associations to the development and implementation of new technology.

3.13 Current agreements between the organisation and its unions and staff associations on the subject of new technology.

3.14 Which groups of staff the company will benefit most from making more efficient and effective.

3.15 Which are the most appropriate areas in the company to introduce new technology to first.

4 *Communication and Information Flows*

4.1 Determine the business Communication Costs by totalling the individual costs of each of the following:

— internal telephone system;

— external telephone facility;

— intercom system;

— bleeper system (paging systems);

— internal telegraphy (telex) system;

— public telex facility;

— facsimile system;

- in-house transport for personnel and communications;
- travel agency service;
- ticket and expense claims;
- internal mail;
- external mail;
- Intelpost costs;
- Datapost costs;
- regular meetings (notional cost of participants' time);
- ad hoc meetings (notional cost of participants' time);
- central typing/word processing centres;
- secretarial typing (notional cost of secretaries' time and equipment cost);
- printing departments;
- addressing and enveloping machines or services;
- photocopying machines or services;
- collating machines or services;
- correspondence registries;
- central filing facilities;
- personal filing (notional cost of people's time and equipment cost).

For each of the above categories a list of shortcomings should be drawn up. In addition the following questions should be asked:

- how critical is it to the business?
- what alternatives exist in the case of a total breakdown?

4.2 What networks does the organisation use?

4.3 What physical routes the networks take and what associated equipment is employed?

4.4 Who has responsibility for each network?

4.5 Communications between groups based on functional activities.

4.6 Communications between groups formed by departments.

4.7 Communications between groups formed by geographic separation.

5 *The Organisation's Systems and Procedures*

5.1 What computer systems are currently in operation, being developed or planned for development? what official procedures are in use in the company?

5.2 Comparison of systems map with other organisation structures already identified, ie functional activities; departmental organisation; geographical location.

5.3 What terminals are currently in use or planned for use in the organisation?

5.4 The experience that departments have had in implementing new technology.

5.5 Which departments are receptive to change? indifferent to change? or resistant to change?

6 *The External Environment*

6.1 Postal and Telecommunication regulations:

- keep an eye out for official statements or documents issued by the Secretary of State;
- find out the views of the political parties on the subject.

6.2 British Telecom and Post Office plans for changes to current services and for the introduction of new services:

- keep an eye open for papers, articles or talks by BT or PO managers and professionals;
- ask your local BT or PO area management;
- join the Telecommunications Managers Association.

6.3 Data protection legislation:

— Contact the following for information:
  a The National Council for Civil Liberties;
  b The British Computer Society;
  c NCC.

— Obtain the Lindop report;

— Obtain the Younger report;

— Obtain the Council of Europe's Convention for the protection of individuals with regard to the automatic processing of personal data.

6.4 Standards for communicating word processors:

— ensure that any new word processors you purchase are able to communicate with those word processors you currently possess;

— establish if any of the word processors you already have installed communicate with each other and if so what had to be done to achieve that communication;

— find out what interpreting message switch products are available on the market and whether they meet your needs. Find out what plans British Telecom has for a black box protocol converter;

— establish what relevant work is currently being done by the British Standards Institution by contacting the NCC Standardisation Office.

6.5 Teletex standards:

— obtain copies of teletex standards;

— find out current activity by contacting the NCC Standardisation Office;

— ask British Telecom about the UK teletex service they are offering;

— ask all the manufacturers you deal with and talk to if they intend to build teletex capabilities into their workstations.

6.6 Open systems interconnection standards:

— read the NCC Publication *Introduction to Open Systems Interconnection:*

— find out what standards work is currently being done in this field by contacting the NCC Standardisation Office.

6.7 Transborder dataflow regulations:

— obtain the privacy and transborder dataflow guidelines drafted by the OECD in 1979;

— obtain the proceedings of the IBI 'transborder dataflow policies' conference in 1980;

— read journal articles on the subject;

— subscribe to the journal *Transnational Data Report*.

7 *EMS products and services*

7.1 Ensure that you have a basic understanding of the following techniques and services:

— conferences;

— formal meetings;

— informal meetings;

— TV;

— slow-scan TV;

— Confravision;

— national and international phone services;

— internal phone services (PABX);

— teleconferencing;

— radio;

— intercom;

— paper;

— notice-boards;

— tape recorders;

— telephone answering machines;

— video tape recorders;

— telephone answering service (TAS);
— Royal Mail;
— Telex;
— Teletex;
— communicating word processors;
— viewdata;
— Prestel;
— microfilm/fiche;
— facsimile;
— video disks;
— screen-based electronic mail;
— computer conferencing.

7.2 Ensure that you understand how the various products and services will all fit together by:

— asking suppliers to tell you how they see the office of the future;
— talking to other organisations about their experiences and expectations;
— joining a formal working group organised by consultancy organisations.

Checklist D
Human Aspects Criteria (VDUs, Keyboards)

SCREENS

Character Formation

a Character height should be greater than or equal to 3 mm

b The resolution of the dot matrix must be a minimum of 5 x 7 and preferably 7 x 9

c Upper- and lower-case character width should be 70 – 80% of the upper-case character height

d Stroke width should be between 12% and 17% of the upper-case character height

e The space between the characters should be between 20% and 50% of the upper-case character height

f Row spacing should be between 100% and 150% of the upper-case character height

g For lower-case characters, descenders should project below the base line of the matrix

h It should be possible to distinguish between the following characters:

Mutual	*One-way*
O and Q	C called G
T and Y	D called B
S and 5	H called M or N
I and L	J, T called I
X and K	K called R
I and 1*	2 called Z*
	B called R, S, or 8*

*These three often comprise 50% or more of the total confusions.

i It should be possible to distinguish between the number 0 and the letter O. (It should be noted that the letter 'Ø' is included in several Nordic alphabets and should not be used to represent the number 0)

j The characters should be upright, not slanted

k If filters are used, the character image should still appear sharp and well-defined

l The displayed character image should be stable

Coding Format

a It should be possible to distinguish between different luminance levels (used for selective brightening) at maximum setting

b A cursor should be provided

c The cursor should be clearly distinguishable from the other symbols on the display

d If it is possible to blink selected parts of the screen the blink rate should be between 2 and 4Hz

e If the cursor has a blink action it should be suppressible

f All displayed symbols should be unambiguous

The Display Screen and Luminance

a The character luminance should be a minimum of 45 cd/m^2 or preferably between 80 and 160 cd/m^2

b Character luminance should be easily adjustable

c Character image should remain sharply defined at maximum character luminance

d Background screen luminance should be between 15 and 20 cd/m^2 under normal office lighting conditions

e The background screen luminance should be adjustable

f The contrast between character and background should be a minimum of 3:1 and preferably 8:1

KEYBOARDS

General Criteria

a The keyboard should be detachable from the screen, ie joined by a cable

b The keyboard should be heavy enough to ensure stability against unintentional movement

c The thickness of the keyboard (ie base to the home row of keys) should be a maximum of 50 mm and preferably 30 mm

d The angle of the keyboard should be in the range 5–15°

e The surface of the keyboard surround should be matt finished

f The reflectance of the whole keyboard surface (not single keys) should be between 0.40 and 0.60

g When keystrokes are no longer being registered, eg when memory is full, a warning should be provided

Key Characteristics

a Key pressure should be between 0.25 and 1.5 Newtons

b Key travel should be between 0.8 and 4.8 mm

c For square keytops the length of the diagonal should be between 12 and 15 mm

d Centre spacing between adjacent keys should be between 18 and 20 mm

e Key legends should be resistant to wear and abrasion, ie the legends should be moulded into the keytop

f The keytop surfaces should be concave

g Reflections from the keytop surfaces should be kept to a minimum

h The activation of each key should be accompanied by a feedback signal such as an audible click, a tactile click or a snap action

j If two keys are activated simultaneously a warning signal should be given

k The keyboard should be provided with an n-key roll-over facility

Keyboard Layout

a The layout of the alpha keys should correspond to that of a conventional typewriter keyboard layout

b The layout of the numeric keys above the alpha keys should correspond to the conventional typewriter keyboard layout

c If an auxiliary numeric keyset is provided, the keys should be arranged in the same way as either the calculator layout (ie 7, 8, 9 along the top row) or preferably the telephone layout (ie 1, 2, 3 along the top row)

d The space bar should be at the bottom of the keyboard

e The colour of the alphanumeric keys should be neutral (eg beige or grey), as opposed to black, white or one of the spectral colours such as red, yellow, green or blue

f The different function key blocks should be distinct from the other keys by either colour, shape, or position

g All keys for which unintentional or accidental operation may have serious consequences should be especially secure by either their position, a higher required key pressure, a key lock or by a two-handed or two-key chord operation

h The function key labels and symbols should correspond to the same functions on other keyboards in the same workplace

Index

adaptation, of systems	45-46
Alvey Programme	120
AUGMENT (Tymshare) system	28
automation	14, 15-17
(see also office automation)	
BLAISE	28
Canada	104
checklists	159-196
computer-assisted learning	98
conferencing	27-28
Confravision service	27, 28
Consultants in Information Technology (company)	120
consultants, use of	39-40
contracts	43, 83-85
costs	14-15, 75-77
d'Agapeyeff, Alex	120
design	
— of dialogues	70
— of jobs	69
— of systems	40-42, 63-72
— of workplace	71-72
de-skilling	97

electronic mail 25-27
employment 90-92
ergonomics 108-110
Europe 16, 30
European Space Agency 28
expert systems 115, 116, 119-120
eyestrain 106-107

facsimile 29-30
feasibility study 37, 40, 64-67, 73
Financial Advisor (expert system) 120
Ford, Henry 16
France 106
future trends 115-120

health 103-108
Helix (company) 120
HUSAT (Loughborough University) 89

IBM Personal Computer 120
implementation, of office systems 44-45
informal factors 57-58
information
— access 28-29
— as a resource 14-15
— centres 60
— management 59-60
insurance 85
integrated systems 117-118
Investment Advisor (expert system) 120
Isolink telecommunications 120

Jacquard (company) 25
Japan 16, 30, 45

Lloyds of London 85
local area networks (LANs) 31-32
Lyons 13
Lyons Electronic Office (LEO) 13

mailbox
(see electronic mail)
management 53-59, 89-90
management structures 54-55
managers, middle 56-57
manufacturers
(see suppliers)
matrix management 54
Metropolis (film) 16
morale (and motivation) 96-97

National Computing Centre 24, 45, 79, 80
networks 31-32
Norway 103-104

ODEFAS (facsimile) 30
office automation
— activities 18-20
— benefits 17-18
— equipment 20-21
— facilities 23-32
— milestones 118-119
— objectives 38-40
office technology projects 61-62
optical character recognition (OCR) 30, 32

PABX facilities 32
posture 108
Prestel 21, 23, 29
proposal evaluation 81-83

requirements specification 42, 67, 73

safety 104-105
Scandinavia 92
(see also Norway)
SDC Search Service 28-29
security 45
selection, of office systems 73-86

staff 40-41, 43-44, 52, 69, 70, 71, 89-92
strategic framework 37-38
strategy for change 35-47
suppliers 42-43, 67, 73, 77-79, 80-81, 116

Taylor, Frederick Winslow 16
Taylorism (scientific management) 16, 17
technology agreements 40-41, 92-96
teleconferencing
(see conferencing)
Telex 24, 32
tender
— document 80-81
— going to 73, 79-81
— invitation to 43
— sample document 143-148
Toronto 104
Toshiba (company) 30
Trades Union Congress (TUC) 93-94
trades unions 40-41, 93-96
training 97-100
transition, from old to new systems 43-44, 51-62
turnkey services 77

United States 16, 29, 30

vendors, of office systems
(see suppliers)
Visicalc 32
visual display units (VDUs) 103-104, 105-107

wide-area networks 31
word processing 24-25
workstations 18, 32, 71, 120

Xibus workstation 120